Literatures, Cultures, and the Environment

Series Editor
Ursula K. Heise
University of California
Department of English
Los Angeles, CA, USA

Literatures, Cultures, and the Environment focuses on new research in the Environmental Humanities, particularly work with a rhetorical or literary dimension. Books in this series explore how ideas of nature and environmental concerns are expressed in different cultural contexts and at different historical moments. They investigate how cultural assumptions and practices, as well as social structures and institutions, shape conceptions of nature, the natural, species boundaries, uses of plants, animals and natural resources, the human body in its environmental dimensions, environmental health and illness, and relations between nature and technology. In turn, the series makes visible how concepts of nature and forms of environmentalist thought and representation arise from the confluence of a community's ecological and social conditions with its cultural assumptions, perceptions, and institutions.

More information about this series at
http://www.palgrave.com/gp/series/14818

Lance Newman

The Literary Heritage of the Environmental Justice Movement

Landscapes of Revolution in Transatlantic
Romanticism

palgrave
macmillan

Lance Newman
Westminster College
Salt Lake City, UT, USA

Literatures, Cultures, and the Environment
ISBN 978-3-030-14571-2   ISBN 978-3-030-14572-9   (eBook)
https://doi.org/10.1007/978-3-030-14572-9

Cover image: Everett Collection Historical / Alamy Stock Photo

This Palgrave Macmillan imprint is published by the registered company Springer Nature Switzerland AG.
The registered company address is: Gewerbestrasse 11, 6330 Cham, Switzerland

CONTENTS

CHAPTER 1

Landscapes of Revolution

This book sets out to make the literature of the Age of Revolutions into a useful tool for the global environmental justice movement. It answers a call issued by Kathleen Dean Moore and Scott Slovic, who urge us to "feel the heat" of "the emergency of global warming" and to "do the work of the moment." While many previous studies have shown how mainstream ways of writing about nature during the Transatlantic Romantic era helped justify colonialism and imperialism, few have examined the period's alternative currents of environmental literature, which often posed radical critiques of modernity and even argued that another world is possible. A few specialists have begun to study unique ways of writing about nature in early literature by African American, Native American, working-class, and women writers. But until a decade ago, most of the relevant texts were locked up in research libraries, so conversations about them were restricted to professional literary historians. Now, they are readily available to all readers in online collections like Google Books, the Internet Archive, and HathiTrust. This book showcases materials from these archives in order to demonstrate their continuing value. It addresses not only literary scholars, but also environmental justice activists and others who seek inspiration in literature. It maps a literary heritage—a usable past—that can inform and inspire the work of our moment.[1] Transatlantic Romantic environmental literature can awaken us not only to the intricate connections between human and natural systems, but also to the complexity of the threats they

1

L. Newman, *The Literary Heritage of the Environmental Justice Movement*, Literatures, Cultures, and the Environment, https://doi.org/10.1007/978-3-030-14572-9_1

face and to the character of the response that we must mount. The ways that Romantic writers moved their readers to work for change can help us imagine ways to move people to do the same now.[2]

LITERATURE AND ENVIRONMENTAL JUSTICE

As capitalism has spread around the globe during the last three centuries, there have been steep increases in human environmental impacts, such as depletion and pollution of soil and water, deforestation and desertification, and the rapid accumulation of atmospheric carbon dioxide. Throughout this period, literature has played a crucial role in the movements that have emerged to protect nature. First, environmental literature has helped us learn to see places that deserve our appreciation and protection. For instance, Thomas West's *A Guide to the Lakes* (1778) focused initial public awareness on the previously little-known mountains of northwest England.[3] The poetry of William Wordsworth, Samuel Taylor Coleridge, and Robert Southey stimulated further interest. In his *Guide through the District of the Lakes* (1810), Wordsworth wrote that the region should be regarded as a "sort of national property in which every man has a right and interest who has an eye to perceive and a heart to enjoy."[4] Eventually, sustained literary attention transformed the damp and difficult mountains of Cumberland, Westmoreland, and Lancashire into a landscape of superlative natural beauty with a new and much grander identity, the Lake District. As a result, tourists began to rival sheep as a source of profit, especially after the arrival of the railroad in 1847 made the area easily accessible. Conservation efforts began soon after it was clear that intensive visitation was impacting the region's scenic character. After a century of mostly informal management, the Lake District National Park was established in 1951 in order to "conserve and enhance the natural beauty, wildlife and cultural heritage" of what had now become a treasured landscape, a symbol of the English nation whose cultural and economic value depended, at least in part, on the health of its carefully administered natural systems.[5]

At times, the role of environmental literature in preservation efforts has been less direct. For instance, Henry David Thoreau wrote lovingly about the natural community around Concord, Massachusetts in the 1850s and even made some early statements in favor of public land management and the preservation of wilderness for recreation and study. But his books sold poorly and his ideas were mainly ignored during his lifetime. Then, in the

1870s, John Muir became an enthusiastic reader of Thoreau when he lived in the Yosemite Valley. Muir experimented with nature writing as a tool for protecting wilderness from mining, logging, and grazing. His essays and newspaper articles helped secure the conversion of President Abraham Lincoln's Yosemite Grant into a national park in 1890. When Muir founded the Sierra Club two years later, he famously took its motto, "In wildness is the preservation of the world," from Thoreau's essay "Walking."[6] Later, the Sierra Club, under the leadership of David Brower, published coffee-table books that combined Thoreauvian nature writing with landscape photography by figures like Ansel Adams and Eliot Porter. These books proved to be powerful tools for mustering public support during the campaigns against dams at Echo Park, Marble Canyon, and Bridge Canyon.

In addition to helping to protect particular wild places, environmental literature has changed public attitudes toward whole regions. For instance, the Desert Southwest in the United States was transformed from a national sacrifice zone into a signature landscape by writers like Mary Austin, John C. Van Dyke, and Joseph Wood Krutch. On an even larger scale, environmental literature has changed our most basic ways of thinking about the nonhuman world as a whole and our place in it. For instance, Aldo Leopold's *Sand County Almanac* proposed a land ethic that "enlarges the boundaries of the community to include soils, waters, plants, and animals." Our membership in this expanded community implies that we all carry "individual responsibility for the health of land."[7] Leopold's ideas became widely influential during the second half of the twentieth century and provided the philosophical underpinning for much of the environmental movement's work during that crucial period of grand legislation and large-scale preservation.

Environmental literature has also helped us recognize new environmental threats. For example, Rachel Carson's celebrated exposé, *Silent Spring*, with its haunting description of a town where birds have been poisoned into silence, alerted the world to the dangers of the toxic chemical DDT. In addition to informing its readers of this new threat, *Silent Spring* motivated them to take action. Carson is widely credited with moving hundreds of thousands of people to participate in the wave of environmental organizing that culminated in the first Earth Day on April 22, 1970. That watershed event was partly responsible for the creation of the Environmental Protection Agency in December of the same year and the Occupational Safety and Health Administration the next year. In 1972, Congress sent the Clean Water Act and the Marine Mammal Protection

Act to President Richard Nixon. A year later, the Endangered Species Act arrived in his office. The proverbially conservative Nixon signed each of these progressive bills into law because, at a time when millions of people were rallying to save the Earth, the political costs of refusing to approve them would have been far too high.

New threats sometimes require new responses, and environmental literature has inspired some readers to adopt more militant stances and to employ more forceful tactics in defense of nature. Not long after Nixon resigned, Edward Abbey's novel *The Monkey Wrench Gang* (1975), with its unabashed celebration of direct action to stop the destruction of wilderness by industrial development, inspired a generation to engage in solitary acts of "night work" or eco-sabotage. Many of the same activists participated in daylight campaigns of direct action, such as the long-running defense of California's old-growth redwoods in which activists used their own bodies to block logging roads and to protect individual trees from chainsaws. More recently, in January 2013, Michael Brune, the Sierra Club's executive director, explained that his organization had decided to adopt more radical political tactics than ever before in response to the oil industry's refusal to reduce carbon dioxide emissions:

> If you could do it nonstop, it would take you six days to walk from Henry David Thoreau's Walden Pond to President Barack Obama's White House. For the Sierra Club, that journey has taken much longer. For 120 years, we have remained committed to using every "lawful means" to achieve our objectives. Now, for the first time in our history, we are prepared to go further. … Next month, the Sierra Club will officially participate in an act of peaceful civil resistance. We'll be following in the hallowed footsteps of Thoreau, who first articulated the principles of civil disobedience 44 years before John Muir founded the Sierra Club.[8]

The issue that finally motivated the Sierra Club to align itself with Thoreauvian militancy was the Keystone XL pipeline that is designed to connect tar sands production facilities in Alberta to refineries and ports in the United States. Since Bill McKibben focused international attention on this formerly obscure oil infrastructure project, it has been an *cause célèbre* for climate activists who hope to escalate the fight against global warming. They recognize that reducing modern society's carbon emissions will require radical changes in the way that extremely powerful energy and manufacturing corporations do business. Not surprisingly, those companies have invested heavily in campaigns of scientific disinformation and political lobbying to protect their interests. As a result, the federal government has been slow and timid in its response to an emergent problem that the scientific community has clearly understood and publicized for at least two decades. In this situation, the Sierra Club's leaders recognized that more confrontational methods were required. To find them, Brune and others reread the literary touchstones that had informed their organization for more than a century.

Environmental threats are much more intractable and complex now than they were 120 years ago, or even 50 years ago. For instance, commercial logging, which once caused local or regional loss of habitat and soil, has now evolved into the global problem of massive deforestation, which contributes both to climate change and to the runaway loss of biodiversity. Water pollution, which once affected individual streams, rivers, ponds, and lakes, now takes the form of granular plastic dispersed throughout the world's oceans. And of course, air pollution, which once meant photochemical smog in local or regional airsheds, now takes the form of atmospheric carbon dioxide concentrations above 400 parts per million. In addition to the spread of familiar problems, we now face new ones, like widespread radioactive contamination from expended munitions and nuclear accidents, antibiotic-resistant strains of disease-causing microorganisms, and contamination of our food and water supplies with pharmaceutical drugs.

Just as environmental threats have grown in scale and complexity, so have the corporations and governmental institutions that bear the heaviest responsibility for them. The City Bank of New York, founded in 1812, is now Citigroup, which controls almost $2 trillion in capital and operates in 112 countries. Standard Oil, which was founded in Ohio in 1870, is now ExxonMobil Corporation, which operates in almost every country on the planet and generates an annual revenue of more than $400 billion. Dow Chemical, which started as a Michigan bleach factory in 1897, now operates around the world, manufacturing and selling plastics and chemicals mainly for sale to other businesses. The global business environment in which these corporations operate is managed by powerful financial institutions like the World Bank and the International Monetary Fund, and it is secured by the U.S. military, which now maintains permanent bases in more than 100 countries.

Because globalization has intensified environmental threats and dispersed them so widely, the number of human communities and ecosystems impacted by them has been vastly multiplied. A planetary perspective reveals that environmental issues cannot be separated from the social, polit-ical, economic, and cultural frameworks within which they take shape. In a world structured by international capitalism and the long history of colo-nialism, environmental benefits and burdens are unevenly distributed both within and between nations. For instance, toxicity and pollution have reshaped the landscape of illness and mortality. The major biological dis-eases of the past, like tuberculosis, influenza, polio, and smallpox, have been mainly eradicated. But they have been replaced with epidemics of new environmental disorders, such as cancer, asthma, allergies, heart disease, obesity, and infertility. Environmental threats are not simply *out there* in nature. They are also *in* our own communities and our bodies. Particulate air pollution from automotive and industrial sources travels across borders and even oceans, where it is inhaled into human and animal bodies, crosses the membranes in the lungs, enters the blood stream, and accumulates in the vital organs, including the heart and brain. In our global system, the rich produce environmental impacts and the poor experience them.

Despite these changes in the nature of environmental threats, as well as in the character of the communities at risk, large sections of the main-stream environmental movement have been quite slow to adapt. As soci-ologist Dorceta Taylor argues, a very narrow band of concerns has tended to dominate most discussions of the environment for the last two centu-ries, and the time to change the subject has long passed:

> The history of American environmentalism presented by most authors is generally limited to the perspective of White middle-class male environmen-tal activism. The tendency to view all environmental activism through this lens limits our understanding of how class, race, and gender relations struc-tured environmental experiences and responses over time …. The environ-mental movement is a powerful social movement; however, it faces many challenges. Among the most urgent is the need to develop a more inclusive, culturally sensitive, broad-based environmental agenda that will appeal to many people and unite many sectors of the movement.[9]

The issue of diversity in the environmental movement has been on the table for decades, at least since the Southwest Organizing Project's 1990 letter to the Group of Ten environmental organizations, which denounced the mainstream movement's failure to integrate the perspectives of people of color and called on them to diversify their leadership.[10] There have been significant efforts since then, and Taylor acknowledges that those efforts have led to "significant progress on gender diversity." However, she observes that, despite "being more than 30% of the US population and supporting environmental protections at higher rates than whites, people of color have not broken the 16% 'green ceiling' in any of the Environmental Organizations surveyed" for the July 2014 Green 2.0 report.[11]

A grassroots global environmental justice movement has grown rapidly in response to threats and conflicts that mainstream organizations have failed to address because of their cultural exclusivity. There are now hundreds of thousands of environmental organizations around the world, ranging from international NGOs to issue-focused nonprofits to tribal networks to neighborhood campaigns. As Paul Hawken observes in *Blessed Unrest* (2007), the many millions of people, most of whom are people of color, who actively work to protect and restore the environment comprise the largest movement of any kind in the world. The movement is not just large, but also powerful, "capable of bringing down governments, companies, and leaders through witnessing, informing, and massing." It is also decentralized and fluid, capable of evolving rapidly in response to new threats and opportunities. Hawken observes that the new global environmental movement "has three basic roots: environmental activism, social justice initiatives, and indigenous culture's resistance to globalization, all of which have become intertwined."[12] There is a tremendously important opportunity at hand for environmentalists whose experience is rooted in traditional conservation work to learn from the world's poor that struggles to protect nature and to achieve justice go hand in hand.

Fortunately, because we now have access to so much more information and more powerful communication tools than were available just a few decades ago, new shared interests have become visible and new networks of cooperation are emerging.[13] As Kimberly Ruffin observes in *Black on Earth*, "Human biodiversity now stands to enrich cross-cultural dialogue about environmentalism in ways it has never done before."[14] Indeed, fostering such dialogue will be our most important work in the coming decades.[15] In order to do that work well, we will need new literary touchstones that can help us imagine and belong to a truly diverse activist community. We need to look for new kinds of texts by authors who have not previously received the attention they deserve, texts by the women, people of color, and workers whose communities, workplaces, and bodies have been jammed by the daily traffic of cars, trucks, planes, pharmaceuticals, pesticides, plastics, memes, and images.[16] In order to recognize such texts when we see them, we need to adopt a broader definition of environmental literature, one that, as Joni Adamson writes, will allow us to "search out the most thoughtfully imaginative works of environmental reflection – not only those that confirm preconceived notions about nature, but also those that challenge perceptions and expand definitions."[17] Most importantly, we need to learn to recognize the environmental significance of literary texts that we have previously understood as being exclusively concerned with social justice.[18] This reorientation is needed in our reading of not just contemporary literature but historical literary texts as well, particularly those written during the Romantic century, when capitalism and environmentalism spread together around the Atlantic Rim and the across the planet.[19]

THE RADICAL PASTORAL AND THE REVOLUTIONARY SUBLIME

Soon after Ralph Waldo Emerson's *Nature* was published in 1836, his friend Christopher Pearse Cranch moved from Massachusetts to Kentucky to serve as editor of the Louisville-based Unitarian magazine *The Western Messenger*. From his Western outpost, Cranch followed the ferment of ideas that stirred Emerson and his circle during the volatile early days of Transcendentalism. A year or two later, he created a series of dry-witted sketches that he bound together and called *Illustrations of the New Philosophy*. One is keyed to the most famous passage in *Nature*, Emerson's hallucinatory experience of transcendence: "Standing on the bare ground, my head bathed by the blithe air, and uplifted into infinite space, all mean egotism vanishes. I become a transparent eyeball."[20] In Cranch's sketch, Emerson's body is reduced to a pair of stick-like legs that are covered by a prim waistcoat and cravat.[21] This attenuated body's sole function is to carry an enormous eye, which serenely contemplates a sky that is empty of all but a few wisps of cloud. The stately, spiritualized Emerson traverses a soft-focus New England with low, rounded hills in the background. The surface of the earth is so perfectly smooth that he can walk on tender bare feet. The sketch's perspective places the viewer's eye at the same elevation as Emerson's, so that, like him, we see no human activity on the land, except for the spires of a village that peek over a ridge in the middle distance. The passage continues in breathless staccato: "I become a transparent eyeball; I am nothing; I see all; the currents of the Universal Being circulate through me – I am part or particle of God."[22] Emerson the philosopher has dissolved into the divinity of Nature.[23] The price of his ecstatic dematerialization is that he is alienated from the rural hamlet that lies at his feet in Cranch's sketch: "In the wilderness, I have something more connate and dear than in the streets or villages." He has become a "lover of uncontained and immortal beauty," and now human relationships are beneath his notice: "The name of the nearest friend sounds then foreign and accidental: to be brothers, to be acquaintances – master or servant, is then a trifle, and a disturbance."[24]

Of course, in the United States in 1836, the relationship of master to servant was anything but a trifle, whether Emerson chose to pay attention or not. *The Liberator*, the newspaper of the New England Anti-Slavery Society, had been circulating for five years, and the question of human brotherhood was firmly on the table. In turning *to* an idealized and dehumanized Nature, Emerson was turning his attention *away* from the burning issues of liberty and equality that inspired his radical contemporaries. He would later come to support the abolition of slavery, but at this moment in his intellectual career, the moment to which American environmentalists have turned for inspiration for more than a century, he presents a clear choice with a foregone conclusion: You may care about Nature, the pristine space where we make solitary contact with the divine, or you may care about the trivial accidents of the village streets, such as who owns whom, but not both. The time has come for environmentalists to look elsewhere for inspiration and guidance.

One place to start is William Blake's illustrated poem *America: A Prophecy* (1793), which has inspired activists for social justice for more than two centuries, but has rarely been read for its environmental meanings. At the beginning of the poem, the leaders of the American army have gathered on the coast at night. They look toward England, and their faces glow bloodred, lit by "Sullen fires across the Atlantic." On the other side of the ocean, the "Guardian Prince of Albion burns in his nightly tent," enraged by the rebels' audacity. General George Washington declares that "a heavy iron chain/Descends link by link from Albions cliffs across the sea to bind/Brothers & sons of America."[25] A cataclysm of violence follows and a demi-god, Orc, emerges from the bloodred clouds of war. In Blake's portrait of Orc,[26] a naked youth sits on a sun-dappled hilltop singing a hymn to liberty:

> Let the slave grinding at the mill, run out into the field:
> Let him look up into the heavens & laugh in the bright air;
> Let the inchained soul shut up in darkness and in sighing,
> Whose face has never seen a smile in thirty weary years;
> Rise and look out, his chains are loose, his dungeon doors are open.

And let his wife and children return from the opressors scourge;
They look behind at every step & believe it is a dream.
Singing. The Sun has left his blackness, & has found a fresher morning
And the fair Moon rejoices in the clear & cloudless night;
For Empire is no more, and now the Lion & Wolf shall cease.[27]

In Orc's song, oppression takes place in the mills and prisons of imperial Britain, while freedom grows in peopled green fields that are lit cheerfully by the sun and moon. This stirring vision of human equality in an open and healthy countryside is one of the most powerful symbolic landscapes of the Romantic era. It is a version of the radical pastoral *topos*.

Topos, from the Greek τόπος, "place," refers to a traditional literary motif, a rhetorical commonplace or meme, that is both stable and dynamic.[28] *Topoi* are familiar and well-defined "places" of thought where it is nevertheless possible to "find" new ways of thinking through rearrangement, recombination, reinterpretation, and similar processes.[29] While *topos* is a somewhat obscure term, it contains a useful reminder that many of our ways of seeing the world have persisted over long periods of time *and* that they continuously adapt to new historical and literary environments.[30] Blake, for instance, does not simply reproduce the traditional pastoral *topos* that had seen so much service in English poetry from Spenser and Marlowe to Pope and Gray. Instead, he radicalizes it to suit new circumstances. In *America: A Prophecy*, the pastoral, which had previously been used to make the power of the English aristocracy seem natural and inevitable, has now been repurposed to meet the needs of an era of militant democracy.[31]

The word *topos*, with its roots in "place," also reminds us that literary landscapes relate to actual places in dynamic and sometimes complex ways. After all, Orc's song about a slave escaping from bondage is no mere fantasy of return to a green and pleasant English countryside. Even if the word "slave" is being used figuratively to emphasize the notorious unfreedom of British mill workers, it also alludes pointedly to American slavery. In the two decades between the beginning of the American Revolution and Blake's composition of his poem, public debates about the international slave trade had made it all too clear, especially to someone like Blake, that the United States was no utopia. Thus, there is a jarring dissonance between Orc's radical pastoral vision of a democratic rural community and the bitter reality that the American republic was built on the backs of slaves. The poem's insurgent energy is rooted in that discord. *America: A Prophecy* is not a naïve dream. It is a call to arms. With its repeated verb, "let," it attempts to stir the reader to action by using the subjunctive mood to show that another world is possible. Orc, the "lover of wild rebellion," calls on us to help *create* a better future, to transform a world that is manifestly *not* free or green.[32] Doing so will require formidable strength and resolve, and a slower look at Blake's engraving reveals that Orc is an appropriately potent figure. On the two previous pages, he has appeared as a dragon and a serpent. Now he now takes the form of Ares triumphant, reclining on the green earth and staring confidently into the heavens. His musculature is schematic and exaggerated. His feet, magnified by foreshortening, grasp the soil. He is sublimely embodied and rooted. His genitals, resting on the green grass, attract the viewer's attention at the very center of the composition. The spirit of revolution has a material body, is an *earthly* body, with all the powers, vulnerabilities, and beauties that such bodies share.

However, Blake has seated this splendidly physical and powerful being on a grave. A moldering skull reminds us of the way of all flesh. At the bottom of the page, a luxuriant thistle and a collection of rampant reptiles ensure that we do not mistakenly imagine Nature to be purely benign. Orc is silhouetted against a skyscape that is dominated by towering thunderheads. These clouds enfold and isolate him, and when he is viewed within this sublime frame, he seems less like an impassioned youth than a commanding anti-hero. Instead of an unclothed Adam, he is a rebel angel who refuses to cover himself with the leaves that hang within easy reach. He is tranquil for now, but potentially lethal. His insolently naked body stands as a bold sign of the shared corporeal life that grounds his vision of freedom in Nature. Crucially, the sublime reminders of eternity and infinity that enclose him also serve to focus and intensify the force of his call for a new stage in the ongoing revolution that has left far too many still in chains.

This second *topos* embedded in Blake's engraving, the revolutionary sublime, evokes the feeling of awe—a combination of admiration and fear—that we so often feel when we contemplate world-changing political, social, and economic upheavals. On the one hand, the revolutionary sublime opens up critical space between readers and disturbing figures like Orc. Locating such characters in wild landscapes that have been purified of social context isolates them so that we can coolly evaluate the demands they make on us. At the same time, framing such revolutionary energies and ideas in this way distills them to their essence and thus makes them all the more clear and compelling.

Blake's engraving, in which these two *topoi* coexist in tension with one another, is an apt emblem of Transatlantic Romanticism, the international cultural movement that evolved during what has been called the Age of Revolutions. On the one hand, the radical pastoral *topos* voiced by Orc was the motive force that drove Transatlantic Romanticism forward. It shaped the political aspirations of the insurgent bourgeoisie who broke the power of the European aristocracy. Later it inspired the postrevolutionary movements that broadened the horizons of freedom by demanding the abolition of slavery, sovereignty for Native Americans, and equal rights for women and workers. The radical pastoral *topos* also energized many of the most memorable and provocative works of literary art produced during the period. Orc himself was reincarnated again and again as the anti-heroes at the center of such works as William Godwin's *Things as They Are; or The Adventures of Caleb Williams* (1794), William Earle's *Obi; or, The History of Three-Fingered Jack* (1800), Percy Bysshe Shelley's *Prometheus Unbound* (1820), Lydia Maria Child's *Hobomok, A Tale of Early Times* (1824), John Greenleaf Whittier's poem "Toussaint L'Ouverture" in his collection *Voices of Freedom* (1846), and even Walt Whitman's *Leaves of Grass* (1855). On the other hand, the radical pastoral *topos*, along with the movements it inspired, often provoked negative reactions that ranged from fear and anxiety to hostility and aggression. So, many of the texts in which it appears share the kind of internal tension that makes Blake's engraving so compelling: They isolate anti-heroes in revolutionary sublime landscapes that simultaneously contain and amplify their radical pastoral ideas. A closer look at this dynamic tension in two key Romantic novels, Mary Shelley's *Frankenstein* (1818) and Nathaniel Hawthorne's *The Scarlet Letter* (1850), may suggest some of the ways that Romantic *topoi* can become important resources of hope in the present.

THE CREATURE ON THE SUMMIT

When Victor Frankenstein first sees the creature that he has brought to life, he fixates on the spectacle of its marked body: "His yellow skin scarcely covered the work of muscles and arteries beneath; his hair was of a lustrous black, and flowing; his teeth of a pearly whiteness; but these luxuriances only formed a more horrid contrast with … his shrivelled complexion and straight black lips."[33] The creature is abjectly corporeal. But it is not an inert body. It is menacingly vital and self-directed.[34] At the same time, Victor Frankenstein's sharp focus on color racializes the creature. Its body is not white or yellow or black; it is all of these at once. The creature is a kind of composite or hybrid Other, all the more terrifying because of its racial indeterminacy.[35] Frankenstein flees in horror.

At their next encounter, Frankenstein again responds with revulsion: "Its unearthly ugliness rendered it almost too horrible for human eyes." The creature's first words express weary familiarity with the alienation to which it is doomed by its marked body: "'I expected this reception,' said the daemon. 'All men hate the wretched.'" Nevertheless, the creature attempts to persuade its maker to acknowledge that the two of them are "bound by ties only dissoluble by the annihilation of one of us." It calls on Frankenstein to fulfill his paternal obligations: "Do your duty towards me, and I will do mine towards you and the rest of mankind." When Frankenstein instead attempts to murder the creature, it easily evades him. Finally, he renounces the creature, and the terms of his renunciation are quite specific: "Begone! I will not hear you. There can be no community between you and me." And again: "Begone! Relieve me from the sight of your detested form!"[36] The creature, then, is exiled from the human community because of the visual appearance of its body.

After leaving Frankenstein's rooms in a frenzy, the creature first regains awareness of itself and its surroundings when it awakens near a brook in a forest. It imitates the singing of forest birds and learns to kindle fire and forage for food. Its first loss of innocence comes when it wanders into a rural village where the people band together and assault it with clubs and stones. Terror-stricken, it hides in a shed attached to the side of a cottage in the countryside. It spends the following winter in this "kennel," surreptitiously watching an old man, De Lacey, along with his children, Agatha and Felix.[37] These three live a simple life in their pastoral retreat, and by observing them, the creature learns to speak, to read, and to feel sympathy. It also learns about the realities of poverty and the value of kindness. This political education in human nature continues when the creature listens to Felix read aloud from *Les Ruines* (1791) by Constantin-Francois Chasseboeuf, Comte de Volney. From Volney's quintessential Enlightenment narrative of the progress of civilization, the creature learns of "the strange system of human society ... [of] the division of property, of immense wealth and squalid poverty; of rank, descent, and noble blood." Moreover, it forms a sense of justice and a clear understanding of the unfair reasons for its own exile: "I possessed no money, no friends, no kind of property. I was, besides, endowed with a figure hideously deformed and loathsome."[38] The creature imbibes this vision of human equality and achieves political self-consciousness during its immersion in an inset radical pastoral *topos*. Eventually, because it, like all beings, "require[s] kindness and sympathy," it reveals itself to the De Laceys, in the hope that they will "compassionate" it and "overlook [its] personal deformity."[39] Instead, "overcome by pain and anguish," they panic and attack as it cowers on the floor.[40] The creature's anguish is quickly replaced by rage, and it becomes a revolutionary, turning to violence to force open and reconfigure the network of "relationships which bind one human being to another in mutual bonds."[41] First it kills William, Frankenstein's younger brother, and then it frames Justine Moritz, the family housekeeper, who is found guilty of the murder and executed.

The Frankensteins attempt to recover from their losses by making an excursion to the type locality of the sublime, the valley of Chamonix. Frankenstein's description of the place sounds like a parody of the popular Romantic practice of landscape appreciation, with its minute differentiations between multiple categories of scenery: "We passed the bridge of Pelissier, where the ravine, which the river forms, opened before us, and we began to ascend the mountain that overhangs it …. The valley is more wonderful and sublime, but not so beautiful and picturesque as that of Servox." The next day, he goes "alone to the summit of Montanvert" where he sits "upon the rock that overlooks the sea of ice."[42] His surroundings are obscured by clouds above which only he rises. According to the conventional formulas that he has been so self-consciously rehearsing, Frankenstein can travel no farther from the filthy world of Geneva and his laboratory. But instead of finding solace in sublime Nature, he finds the creature. And here, in this otherworldly place, instead of killing its maker, it *speaks*.

The creature delivers a personal narrative that amounts to an impassioned vindication of its natural rights and a militant critique of "the barbarity of man." Since it fled the De Laceys' hut, it has been fated to wander the earth beyond the boundaries of human society: "The desert mountains and dreary glaciers are my refuge. I have wandered here many days; the caves of ice, which I only do not fear, are a dwelling to me, and the only one which man does not grudge. These bleak skies I hail, for they are kinder to me than your fellow-beings." Not surprisingly, the creature has been embittered by the experience of being cast out into the wild because of its physical appearance: "Believe me, Frankenstein: I was benevolent; my soul glowed with love and humanity: but am I not alone, miserably alone? You, my creator, abhor me; what hope can I gather from your fellow-creatures, who owe me nothing? they spurn and hate me."[43]

On the one hand, staging this monologue on the summit of a glaciated mountain emphasizes the dizzying radicalism of the creature's ideas, making them seem extreme and even inhuman. On the other hand, the creature's narrative of awakening shows that these are the simple teachings of Nature. By association, the mountaintop scene adds force to the argument that the creature's "vices are the children of an enforced solitude."[44] In other words, the revolutionary sublime simultaneously quarantines *and* intensifies the creature's claim to human nature and human rights. This double effect only becomes stronger as the narrative stages subsequent encounters between Frankenstein and the creature in increasingly remote and hostile settings, ending at last on the Arctic ice. The novel attempts to send the creature into exile only to find that it has created a martyr.

Frankenstein also employs a second strategy to contain the creature's uncontainable radicalism: it encloses the creature's life narrative within three sets of quotation marks. The creature tells its story to Victor Frankenstein, who then tells it to Robert Walton, who later transcribes it from memory and sends the resulting manuscript to his sister in England, where it has made its way to the press. But instead of removing the creature from the realm of readerly sympathy, this elaborate framing strategy compels us to think through questions of the authenticity of the text and the reliability of its many narrators. It enjoins us to evaluate events from multiple perspectives and to engage in critical compassion. In other words, it levels the field by placing the creature and its maker on the same kaleidoscopic stage and then forcing us to adjudicate as they debate the meaning of their reciprocal acts of violence. The final effect of the novel's elaborate staging is to encourage us to reject Victor Frankenstein's callow arrogance, vanity, and avarice, and to identify instead with the creature's story of learning to sympathize with others and of being driven to violence by prejudice. In the end, *Frankenstein* compels us to agree with the creature's radical pastoral claim that Nature establishes a standard of human rights that we must use to gauge social justice.

THE WOMAN IN THE FOREST

Like Frankenstein's creature, Hester Prynne first appears in *The Scarlet Letter* as a visual spectacle. The Salem magistrates compel her to stand on the pillory before "the public gaze" of the assembled villagers. She is physically imposing, "with a figure of perfect elegance, on a large scale ... characterized by a certain state and dignity."[45] Her majestic body itself is not marked; instead, she has been ordered to wear a "token of her shame" attached to her clothing. The scarlet letter has "the effect of a spell, taking her out of the ordinary relations with humanity, and inclosing her in a sphere by herself."[46] The magistrates of Salem "had set a mark upon her, more intolerable to a woman's heart than that which branded the brow of Cain."[47] The curse of Cain figured regularly in the Romantic period's urgent discussions of slavery and abolition, since he was believed by some Christians to have been marked by God with black skin for his fratricide. Apologists for slavery extended this interpretation to argue that Africans were descendants of Cain and that enslavement was divine punishment for their forefather's crime. Prynne, on the other hand, has inherited another kind of bondage that is especially galling to her "woman's heart," subordination to men in expiation of Eve's misconduct in the Garden. As punishment for defying male authority and fulfilling her bodily desire for Arthur Dimmesdale, she has been marked as guilty of the definitive female transgression, adultery. Rather than meekly accept her sentence, she has defiantly fashioned the letter A out of "fine red cloth, surrounded with an elaborate embroidery and fantastic flourishes of gold thread." When the people of Salem see her, her ostentatiously marked body inspires "horrible repugnance" and "bitterest scorn."[48]

After Prynne's "term of confinement" in prison ends, she starts a new life just beyond the borders of Salem: "Her sin, her ignominy, were the roots which she had struck into the soil."[49] She moves into "a small thatched cottage [that] stood on the shore, looking across a basin of the sea at the forest-covered hills." On this salt-water farm, a stock locale of the New England pastoral, she supports herself and her daughter, Pearl, by "plying her needle at the cottage-window."[50] Prynne lives in this rural retreat for seven uneventful years, and during her "long seclusion from society," she is transformed into a revolutionary. Despite her isolation, her labor remains vital to the community that has "banished" her: Her "needle-work was seen on the ruff of the Governor; military men wore it on their scarfs, and the minister on his band; it decked the baby's little cap; it was shut up, to be mildewed and moulder away, in the coffins of the dead." What she learns from this intimate vantage point makes her feel all the more estranged: "In all her intercourse with society ... there was nothing that made her feel as if she belonged to it."[51] She responds to her sense of alienation by vowing not "to measure her ideas of right and wrong by any standard external to herself."[52] And she transforms herself into a "self-ordained Sister of Mercy" who acknowledges "her sisterhood with the race of man" by giving to the poor and by comforting the sick and dying.[53] More than simply engage in practical acts of sympathy, she turns "from passion and feeling, to thought" and swims in the philosophical current of radicalism that runs from the Reformation to the Enlightenment and beyond:

> The world's law was no law for her mind. It was an age in which the human intellect, newly emancipated ... had overthrown and rearranged – not actually, but within the sphere of theory, which was their most real abode – the whole system of ancient prejudice .... Hester Prynne imbibed this spirit. She assumed a freedom of speculation, then common enough on the other side of the Atlantic, but which our forefathers, had they known of it, would have held to be a deadlier crime than that stigmatized by the scarlet letter.[54]

Of course, the first target of Prynne's speculative reasoning is gender hierarchy. Contemplating "the whole race of womanhood," she asks a rhetorical question, "Was existence worth accepting, even to the happiest among them?" She reaches the feminist conclusion that "the whole system of society is to be torn down, and built up anew. Then, the very nature of the opposite sex, or its long hereditary habit, which has become like nature, is to be essentially modified, before woman can be allowed to assume what seems a fair and suitable position" in society.[55] The novel begins rising to its climax when she discovers that her former husband, the physician or "leech" Roger Chillingworth, has attached himself to her one-time lover, the Reverend Arthur Dimmesdale, and is slowly killing him. She is now no longer the passive subject of Salem's recriminating gaze, and she makes a startlingly bold decision. She intervenes directly in the relationship between two of the town's most eminent men.

Not surprisingly, the novel attempts to contain such an audacious woman, along with her feminist ideas. Thus, when Prynne speculates about revolutionary changes in the relations between men and women, the narrator interrupts her abruptly with a series of sexist remarks. First, he belittles her as a woman who is foolishly dabbling in a male activity: "A woman never overcomes these problems by any exercise of thought. They are not to be solved, or only in one way. If her heart chance to come uppermost, they vanish." Of course, this passage tells us more about the narrator than it does about Prynne. Next, he calls her sanity into question, using peculiar language to do so: "Hester Prynne, whose heart had lost its regular and healthy throb, wandered without a clew in the dark labyrinth of the mind; now turned aside by an insurmountable precipice; now starting back from a deep chasm. There was wild and ghastly scenery all around her, and a home and comfort nowhere."[56] The image of a solitary figure who is frantically lost in a sublime, mountainous landscape is meant to serve as a metaphor for Prynne's mental instability. But the image just as powerfully captures the narrator's intense anxiety, which has been provoked by staring into the chasm of Prynne's feminism. That anxiety is also reflected in the way he turns to what were in 1850 the dustiest clichés of landscape aesthetics, as well as in the way that his sentences falter as he piles up phrases in his hurry to dismiss the force of her subversive ideas.

The Scarlet Letter also opens up the distance between the reader and Prynne by setting her story two centuries in the past. This chronological displacement allows the novel to more fully explore the implications of her feminist critique of a society that will not tolerate her sexual and intellectual freedom. Hawthorne is even able to bring the novel to a conclusion that voices her dream of a "brighter period" that would "establish the whole relation between man and woman on a surer ground of mutual happiness."[57] On the other hand, this displacement also implies that such freedom can only exist in the realm of historical fable. The narrator is quite explicit about the idea that the novel is a fantasy compounded out of historical materials. The infamous introduction centers on a fictional incident, the narrator's discovery, in the heaps of historical rubbish on the second floor of the Custom House, of the embroidered A along with a brief sketch of Hester Prynne's story written on foolscap by an amateur antiquarian. The narrator (who shares some biographical details with Hawthorne) is carefully ambiguous about the relationship between this imaginary manuscript and the "romance" that he has based on it. On the one hand, he proclaims that "the main facts" of The Scarlet Letter "are authorized and authenticated by the document of Mr. Surveyor Pue." However, he immediately complicates this simple statement:

> I must not be understood as affirming, that, in the dressing up of the tale, and imagining the motives and modes of passion that influenced the characters who figure in it, I have invariably confined myself within the limits of the old Surveyor's half a dozen sheets of foolscap. On the contrary, I have allowed myself, as to such points, nearly or altogether as much license as if the facts had been entirely of my own invention.[58]

What he gives with one hand he takes away with the other. The novel is "authorized and authenticated" by an historical source even though it is "as if" it was entirely invented. Moreover, he tells us that he created this "semblance of a world out of airy matter" at a time when he felt his "intellect dwindling away; or exhaling ... like ether out of a phial."[59] Everything about this narrative, then, is elaborately unreliable. The evasive narrator ostentatiously refuses to validate either the repressive Puritan magistrates or the woman who has refused to recognize their authority.

Finally, *The Scarlet Letter* relies on the revolutionary sublime *topos* to frame Prynne and her radicalism. Her climactic meeting with Dimmesdale takes place in a "primeval forest" that "imaged not amiss the moral wilderness in which she had so long been wandering."[60] Hoping to catch Dimmesdale on one of his "meditative walks" and warn him about Chillingworth's malign influence, she sits down to wait for him on "a luxuriant heap of moss" at the foot of "a gigantic pine" in "a little dell." Prynne contemplates the scene before her: "boulders of granite seemed intent on making a mystery of the course of [a] small brook; fearing, perhaps, that with its never ceasing loquacity, it should whisper tales out of the heart of the old forest whence it flowed."[61] Inspired perhaps by this emblem of human vitality distorted by moralistic repression, Prynne finds the freedom to defend her sexual independence: "What we did had a consecration of its own." Even the narrator can acknowledge that here in the woods, "Arthur Dimmesdale, false to God and man, might be, for one moment, true!"[62] And here Hester briefly convinces Dimmesdale, "thou art free!"[63] In the climactic moment that follows, Prynne casts off the scarlet letter:

> The stigma gone, Hester heaved a long, deep sigh, in which the burden of shame and anguish departed from her spirit. O exquisite relief! She had not known the weight, until she felt the freedom! By another impulse, she took off the formal cap that confined her hair; and down it fell upon her shoulders, dark and rich, with at once a shadow and a light in its abundance, and imparting the charm of softness to her features. There played around her mouth, and beamed out of her eyes, a radiant and tender smile, that seemed gushing from the very heart of womanhood. A crimson flush was glowing on her cheek, that had been long so pale. Her sex, her youth, and the whole richness of her beauty, came back from what men call the irrevocable past, and clustered themselves, with her maiden hope, and a happiness before unknown, within the magic circle of this hour.[64]

Prynne's body, her physical and sexual self, so long suppressed by the letter, suddenly bursts forth, not because she is responding to a male touch or command, but because she has avowed her own self-determination. As she comes vividly back to life, the forest responds in kind:

All at once, as with a sudden smile of heaven, forth burst the sunshine, pour-
ing a very flood into the obscure forest, gladdening each leaf, transmuting
the yellow fallen ones to gold, and gleaming adown the gray trunks of the
solemn trees. Such was the sympathy of Nature – that wild, heathen Nature
of the forest, never subjugated by human law, not illumined by higher
truth – with the bliss of these two spirits![65]

Prynne's revolutionary energies and ideas are powerfully authorized by
the sublime forest's sympathy.

The narrator immediately scrambles to remind us that this scene is an
isolated moment of imaginary freedom. *The Scarlet Letter*'s most crushing
turn comes when Pearl, the "material union" of Prynne and Dimmesdale,
cannot tolerate the sight of her changed mother.[66] Prynne retrieves the A
and reattaches it to her dress, and the novel moves on to its deflationary
conclusion. Nevertheless, despite the narrator's final assertion of control
over his insurgent heroine, what remains with the reader is the vivid image
of Prynne's bodily and intellectual liberation in Nature:

Her intellect and heart had their home, as it were, in desert places, where
she roamed as freely as the wild Indian in his woods. For years past she had
looked from this estranged point of view at human institutions, and whatever priests or legislators had established; criticizing all with hardly more
reverence than the Indian would feel for the clerical band, the judicial robe,
the pillory, the gallows, the fireside, or the church. The tendency of her fate
and fortunes had been to set her free.[67]

Because of the authorizing effect of the revolutionary sublime *topos*, *The
Scarlet Letter* is a feminist novel despite itself. In the end, most readers
dismiss the narrator's false objectivity, reject the magistrates' authoritarian
sexism, endorse Prynne's vision of sexual equality, and identify with her
defiant self-determination.

Frankenstein and *The Scarlet Letter* are such powerful novels because
the revolutionary sublime *topos* so urgently authorizes characters who give
voice to radical pastoral visions of freedom and equality. *Frankenstein*'s
center of gravity is the moment at which the creature first speaks on the
mountaintop, thus revealing its full, but distorted, humanity in a cathedral
of Nature. Likewise, the energetic core of *The Scarlet Letter* is Hester
Prynne's declaration of independence in the wilderness. Both novels stage
these climactic moments of self-disclosure and revolutionary protest in
sublime settings that are designed to contain them, but that finally magnify

their force. As a result, both novels *also* isolate these scenes by using elaborate mechanisms of textual subversion. They create literary spaces where everything is provisional, where the central characters' exhilarating ideas are "authorized and authenticated" at the same time that they are presented as if they were pure fantasy. As a result, the novels give a clear voice to their protagonists but refuse to endorse or reject what they say. Because the novels are divided against themselves, we experience them as debates about the meaning of nature and freedom. Is nature, as Victor Frankenstein and the Puritan magistrates believe, an anarchic otherwhere that we visit to be awed into a renewed awareness of the need for social hierarchy and sustained repression of humanity's vicious nature? Or is it, as the life narratives of Frankenstein's creature and Hester Prynne demonstrate, a schoolroom of rebellion where common experience demonstrates the force of egalitarian political ideas? Because these novels refuse to resolve the tensions at their core, because they estrange us from familiar landscapes of thought, they have the power to teach us to see the world from the standpoint of the other, and even to set ourselves as free as the creature and Hester Prynne.

Landscapes of Revolution in Transatlantic Romanticism

The Literary Heritage of the Environmental Justice Movement traces the coevolution of key landscape *topoi* and revolutionary political ideas in environmental literature written in English around the Atlantic Rim during the long Romantic century, 1767–1867. This unusually defined literary historical period ranges from the years leading up to the American Revolution, through the French Revolution and the European revolutions of 1830 and 1848, to the confederation of British colonies into the Dominion of Canada and, in the United States, the end of the Civil War. Of course, texts written in English during this period are not the only ones that are relevant or interesting for environmentalists today. Similar projects of recovery need to be completed in the archives of other languages, literatures, and periods. However, the writers of this time and place were the first to respond, like canaries in a coal mine, to the rise of modern capitalism, which first took root in England, then spread around the Atlantic Rim and throughout the British Empire, and now structures the lives and environments of almost all people around the globe.[68]

Transatlantic Romantic environmental literature may seem to be an unlikely source of new insights into our current moment. After all, many of the period's most canonical texts take flight from politics and society and enact solitary white male encounters with a feminized nature. And all too often, this kind of evasion is accompanied by authoritarian or paternalistic responses to the literary and political aspirations of women, people of color, workers, and the poor. However, Romantic landscape *topoi* were often used in much more critical ways by writers who participated in (or sympathized with) the period's international movements for the abolition of slavery, as well as for women's rights, native rights, and workers' power. This book showcases neglected texts by Romantic writers of color like Charles Ball, Frederick Douglass, William Apess, and George Copway, and others by women writers like Mary Wollstonecraft and Lydia Maria Child. At the same time, it reinterprets familiar texts by writers like William Blake, Nathaniel Hawthorne, Mary Shelley, John Clare, and Henry Thoreau, showing that we can read these familiar texts in productive new ways once we understand that they were written in conversation with their neglected contemporaries.[69] Greg Garrard observes that environmental literature "may cloud our social vision, or open out a human ecological one; it may help in the marginalization of nature into 'pretty ghettoes' or engender a genuine counter-hegemonic ideology."[70] Romanticism did often feature a conservative reaction to revolutionary politics that took the form of a nostalgic turn to an idealized Nature; at the same time, many of the period's more progressive writers staged urgent social conflicts in stylized landscapes that brought them into sharp focus and authorized radical ideas.[71] By doing so, they revealed new truths about a rapidly changing world. Most importantly, they showed how capitalism produces *both* social inequality and environmental destruction. The archive of Transatlantic Romantic environmental literature remains relevant today because the landscape *topoi* that were reinvented by and for the period's activists for human liberation later became the primary tools that environmental activists relied on during campaigns for conservation throughout the twentieth century, and they are tools that we can reinvent to do the environmental justice work of our moment.

The Literary Heritage of the Environmental Justice Movement engages contemporary debates in the environmental humanities, seeking both to engage and to extend what has been called the "materialist turn" or "new materialism," a theoretical movement that rethinks methods of literary criticism in relation to the physicality of nature. The texts discussed here can now be recognized as examples of ecosocial discourse, in which environmental concerns take their place alongside, rather than in opposition to, issues of human rights and justice. These texts frame intersectional critiques of the capitalist ecosocial order, and they argue that achieving race, gender, and class liberation will require transforming human relations with nature. At a time when the dominant Romantic culture of Nature was in full bloom, these texts repurposed the sublime and pastoral modes in order to protest Indian removal, to advocate for native land claims, to condemn the exploitation of both slave and wage labor, to promote African American and working-class land ownership, to denounce the doctrine of gendered spheres, and to assert women's right of access to public space, including nature. In so doing, they identified the material *and* the discursive connections between capitalism's exploitation of nature for profit and its systematic oppression of women, people of color, workers, and the poor.[72] Moreover, these texts foresaw our contemporary understanding that the raced, gendered, and/or classed body is the site where environmental injustice plays itself out in the most starkly material ways. Thus, they present nature and the human body as both symbolic terrains and as real, material things. Finally, these texts deserve our attention because they can inspire us to experience new kinds of self-transformation. They can teach us that, in our encounters with the land, we can do more than recreate the energies that have been exhausted by the pressures and stresses of life in modernity; we can also find the confidence and commitment that we will need to create a better world.

Notes

1. Ammons and Roy curate a sample of the international literature of environmental justice.
2. I have tried to write sentences that breathe freely and to use as little jargon as possible. Occasionally, I use specialized terms, but I do my best to make them accessible and useful. In places where theoretical framing material would ordinarily appear in literary scholarship, the reader will often find passages that connect the literature of the Romantic era to present-day issues in environmental justice. While this book does join current conversations in literary and cultural theory (particularly the "new materialism" in the environmental humanities), it mainly does so by demonstrating through extended close reading of primary texts that Romantic literature addressed the questions that literary scholars are asking now. Further remarks on the scope and methods of this book can be found in the final

section of this chapter. Also, a brief afterword speaks more directly to those who follow specialized professional conversations in the environmental humanities. Because this book discusses so many writers and topics, it is not feasible to include a comprehensive bibliography of relevant scholarly materials. I have relied heavily on the work of many historians, literary critics, philosophers, political theorists, journalists, and others; however, the main text does not discuss their work in detail. Instead, those who are interested will find acknowledgments of debt and suggestions for further reading in these notes. I have limited myself to citing primary texts and those secondary sources that directly influenced my thinking as I was writing. My thanks to Charlene Avallone, Michael Bennett, Jecca Bowen, Kristen Case, Susie Lan Cassel, Sean Desilets, Bryan Evans, Katy Evans, Rochelle Johnson, Ashley Seitz Kramer, Chris LeCluyse, Galina Malugin, Jeff McCarthy, Lauren McCrady, Jeff Nichols, Brent Olson, Paul Outka, Giancarlo Panagia, Jacob Paul, Natasha Sajé, Dennis Sizemore, Rebecca Solnit, Elizabeth Terzakis, Kevin Van Anglen, Laura Walls, and the many others who have provided helpful feedback during this book's long gestation.

3. West was preceded by Thomas Gray, Thomas Pennant, and others, but his *Guide* was the first book-length description of the Lake District that specifically addressed a touristic audience and sold through multiple editions.

4. Wordsworth, 88.

5. Lake District National Park, n.p.

6. Thoreau, *Collected Essays and Poems*, 239.

7. Leopold, 239 and 258.

8. Brune.

9. Taylor, *Race, Class, Gender, and American Environmentalism*, 40–41.

10. Southwest Organizing Project, n.p.

11. Taylor, *State of Diversity in Environmental Organization*, 3–4.

12. Hawken, 12.

13. For example, see Murphy, *Media Commons*, 117–143.

14. Ruffin, 10.

15. Edwardo Lao Rhodes observes that when mainstream environmentalists in the United States are asked about the whiteness of their movement, they often respond that people of color do not care about nature. In order to evaluate this claim, Rhodes analyzes the results of several surveys of environmental attitudes. He reports that "environmental issues are considered crucial by the minority community, and further, that environmental interest is nearly identical for both minority and majority populations." However, "minorities, notably African Americans, may not rank environmental issues as high on their priority list as the majority population does. Other social issues – crime, drugs, and racial discrimination, among others – appear to be

higher priorities for many members of minority populations." Nevertheless, Rhodes shows that "although environmental issues may occupy a slightly lower relative position within parts of the minority community, the surveys indicate that absolute interest in these issues is comparable to that among whites" (75). In other words, people of color are just as worried about nature as anyone else alive on Earth at the moment, but that concern is integrated into a broader awareness of simultaneous environmental and social crises. Rhodes goes on to argue that low participation by people of color in mainstream environmental organizations is most likely the result of the movement's long-standing "anti-urban ethos that allowed for little consideration of the plight of the poor and minorities" (74). In other words, the real problem is that the mainstream environmental movement has defined its core issues so narrowly that it seems irrelevant from an urban multicultural standpoint, and from that standpoint it is all too obvious that environmental issues are inextricably linked to the questions of social justice that the mainstream movement has for so long avoided.

16. One of the first ecocritical studies to look well beyond the insular canon of Thoreauvian nature writing and to sample the literary history of environmental justice is Lawrence Buell's *Writing for an Endangered World*. Murphy has advocated influentially for expanding the field of environmental literature, beginning with *Literature, Nature, and Other*, and continuing in *Farther Afield* and *Ecocritical Explorations*.

17. Adamson, *American Indian Literature*, 19.

18. Armbruster and Wallace argue that if "ecocriticism limits itself to the study of one genre—the personal narrative of the Anglo-American nature writing tradition—or to one physical landscape—the ostensibly untrammeled American wilderness—it risks seriously misrepresenting the significance of multiple natural and built environments to writers with other ethnic, national, or racial affiliations" (7).

19. In *Brave New Words: How Literature Will Save the Planet*, Elizabeth Ammons challenges what she sees as the pessimism, ironic detachment, and even nihilism of much contemporary literary theory and practice. She calls for committed reading and teaching of the "activist tradition of American literature" (ix), because she believes that literary texts "have transformative power. They play a profound role in the fight for human justice and planetary healing that so many of us recognize as the urgent struggle of our own time. Words on the page reach more than our minds. They call up our feelings. They can move us to act" (172). Moreover, Ammons calls for expanding the discipline's traditional focus on what Kwame Appiah and Henry Louis Gates, Jr. called "the holy trinity of literary criticism" (625). Scholars and teachers should fertilize the rich tradition of work that focuses on race, gender, and class by stirring in ecological matter in order to create a holistic perspective

under the rubric of environmental justice. Finally, Ammons also asks us to look beyond the learning goal of critical thinking and to provide answers to "the question of hope" (30). In order to remain relevant in a century of planetary environmental crisis, she writes, the "liberal arts should be offering practical, useful inspiration to everyone seeking to create a different and better world" (13). In particular, she insists on the continuing importance of multiculturalism and directs attention to the emergent idea of ecosocialism, arguing that the job of literature is "to inspire and fortify people in the collective struggle to achieve social justice and restore the earth" (xi).

20. Emerson, *Nature*, 13.
21. See https://iiif.lib.harvard.edu/manifests/view/drs:11324577$4i
22. Emerson, *Nature*, 13.
23. I follow Neil Evernden's "convention of speaking of 'nature' when referring to the great amorphous mass of otherness that encloaks the planet, and [of] speak[ing] of 'Nature' when referring specifically to the system or model of nature which arose in the West several centuries ago" (xi).
24. Emerson, *Nature*, 13.
25. Blake, plate 5.
26. See http://www.blakearchive.org/copy/america.a?descId=america.a.illbk.08
27. Blake, plate 8.
28. According to Carolyn Miller, a *topos* can serve as a tool for rhetorical invention because it "is a conceptual space without fully specified or specifiable contents; it is a region of productive uncertainty" (141).
29. McKeon traces the history of "commonplaces" in rhetorical theory, where they function both as aids to memory and as spurs to creativity.
30. Caroline Levine observes that literary forms (of which *topoi* are a subset) "travel" and "do political work in particular historical contexts." She argues that forms "can survive across cultures and time periods, sometimes enduring through vast distances of time and space." Moreover, forms "reflect or respond to contemporary political conditions [and] shape what it is possible to think, say, and do in a given context" (4–5).
31. Gifford identifies three main uses of the term pastoral to mean (1) a literary tradition with roots in Greek and Roman poems that feature shepherds speaking about their lives, (2) a broader category of literature that contrasts the country and the city, and (3) a judgment that "the pastoral vision is too simplified and thus an idealisation of the reality of life in the country" (2). He also defines two contemporary modalities: anti-pastoral and post-pastoral. Farrier proposes a third, toxic pastoral, and Sullivan a fourth, dark pastoral. All four of these latter-day variations on the pastoral *topos* afford critical analysis of global ecosocial systems in the Anthropocene era. In *Ecocriticism*, Garrard revisits Gifford's three uses of the term and touches on "American pastoral" in order to set up an extended contrast between

"pastoral ecology" and "postmodern ecology" (58). Hiltner argues that English Renaissance pastoral poetry has been misread by critics who see the countryside functioning as no more than a generic setting for satirical commentary on corruption in London and the royal court. Hiltner views these texts as representing nature through gesture rather than mimesis, and he shows that they addressed an emergent environmental conscious-ness spurred by air pollution, wetlands reclamation, and changing agricul-tural practices in colonized Ireland.

32. Blake, plate 9.
33. Shelley, 43.
34. Sandra Gilbert and Susan Gubar argue that the creature's "shuddering sense of deformity, his nauseating size, his namelessness, and his orphaned motherless isolation link him with [Milton's] Eve and with Eve's double, Sin" (239).
35. See Lee for a historicist interpretation of *Frankenstein* in relation to the discourses of slavery and abolition.
36. Shelley, 83–84.
37. Shelley, 90.
38. Shelley, 103.
39. Shelley, 113–114.
40. Shelley, 117.
41. Shelley, 104.
42. Shelley, 73–74.
43. Shelley, 83–84.
44. Shelley, 119.
45. Hawthorne, *Scarlet Letter*, 66 and 63.
46. Hawthorne, *Scarlet Letter*, 64.
47. Hawthorne, *Scarlet Letter*, 99.
48. Hawthorne, *Scarlet Letter*, 62 and 100.
49. Hawthorne, *Scarlet Letter*, 93 and 94.
50. Hawthorne, *Scarlet Letter*, 95–96.
51. Hawthorne, *Scarlet Letter*, 98. Amireh argues that Hester Prynne's charac-terization as "a representative of independent artisanal laborer" engages a well-established convention of using seamstresses to show how industrial-ization and wage labor had eroded the independence of those who were formerly self-employed (92).
52. Hawthorne, *Scarlet Letter*, 192.
53. Hawthorne, *Scarlet Letter*, 194.
54. Hawthorne, *Scarlet Letter*, 198–199.
55. Hawthorne, *Scarlet Letter*, 200.
56. Hawthorne, *Scarlet Letter*, 200–201.
57. Hawthorne, *Scarlet Letter*, 321.

58. Hawthorne, *Scarlet Letter*, 38–39.
59. Hawthorne, *Scarlet Letter*, 44–45.
60. Hawthorne, *Scarlet Letter*, 222.
61. Hawthorne, *Scarlet Letter*, 225–226.
62. Hawthorne, *Scarlet Letter*, 237–238.
63. Hawthorne, *Scarlet Letter*, 240.
64. Hawthorne, *Scarlet Letter*, 247.
65. Hawthorne, *Scarlet Letter*, 247–248.
66. Hawthorne, *Scarlet Letter*, 252.
67. Hawthorne, *Scarlet Letter*, 243–244.
68. McKusick, *Green Writing*, was the first transatlantic study of environmental literature in the Romantic era. It located the origins of "ecological consciousness" in canonical texts by the English Romantics and traced its subsequent development in the writings of the American Transcendentalists. See the introduction of Hutchings, *Romantic Ecologies*, for a remarkably clear framing of the literary critical field of transatlantic postcolonial Green Romanticism (3–32).
69. In an insightful set of remarks on intercultural literary criticism, Joshua Bellin "sees texts as formed by, forming, and a form of encounter: fertile, contested, and multiply determined, they exist on the shifting borders, or in the indefinite field, between peoples in contact" (*Demon*, 6).
70. Garrard, "Radical Pastoral," 464.
71. Buell, "Pastoral Ideology Reappraised," usefully insists on the "ideological multi-valence" of the pastoral mode (21).
72. In *Bodily Natures*, Stacy Alaimo writes, "Casting racism as environmental exposes how sociopolitical forces generate landscapes that infiltrate human bodies .... [T]he penetrating physiological effects of class (and racial) oppression [demonstrate] that the biological and the social cannot be considered separate spheres" (28). Alaimo's new materialist approach to the entangled processes of exploitation and oppression makes it possible to "rewrite the entire expanse of the history of the United States from an environmental justice perspective" (29).

BIBLIOGRAPHY

Abbey, Edward. 1975. *The Monkey Wrench Gang*. Philadelphia: Lippincott.
Adamson, Joni. 2001. *American Indian Literature, Environmental Justice, and Ecocriticism: The Middle Place*. Tucson: University of Arizona Press.
Alaimo, Stacy. 2010. *Bodily Natures: Science, Environment, and the Material Self*. Bloomington: Indiana University Press.
Amireh, Amal. 2000. *The Factory Girl and the Seamstress: Imagining Gender and Class in Nineteenth Century American Fiction*. New York: Garland.

Ammons, Elizabeth. 2010. *Brave New Words: How Literature Will Save the Planet.* Iowa City: University of Iowa Press.

Ammons, Elizabeth, and Modhumita Roy. 2015. *Sharing the Earth: An International Environmental Justice Reader.* Athens: University of Georgia Press.

Armbruster, Karla, and Kathleen R. Wallace, eds. 2001. *Beyond Nature Writing: Expanding the Boundaries of Ecocriticism.* Charlottesville: University Press of Virginia.

Bellin, Joshua David. 2001. *The Demon of the Continent: Indians and the Shaping of American Literature.* Philadelphia: University of Pennsylvania Press.

Blake, William. 1998. *America a Prophecy,* Copy A: Electronic edition, ed. Morris Eaves, Robert Essick, and Joseph Viscomi. Charlottesville: The William Blake Archive.

Brune, Michael. 2013. From Walden to the White House. *Sierra Club.* January 22. https://www.sierraclub.org/michael-brune/2013/01/walden-white-house

Buell, Lawrence. 1989. Pastoral Ideology Reappraised. *American Literary History* 1 (1): 1–29.

———. 2001. *Writing for an Endangered World: Literature, Culture, and Environment in the U.S. and Beyond.* Cambridge, MA: The Belknap Press of Harvard University Press.

Carson, Rachel. 1962. *Silent Spring.* Boston: Houghton Mifflin.

Chasseboeuf, Constantin-Francois, Comte de Volney. 1791. *Les Ruines, ou meditations sur les revolutions des empires.* Paris: Gallica.

Child, Lydia Maria. 1824. *Hobomok, A Tale of Early Times.* Boston: Cummings, Hilliard, and Co. *HathiTrust.*

Cranch, Christopher Pearse. 1837–1839. *Illustrations of the New Philosophy.* MS Am 1506. Houghton Library, Harvard University.

Earle, William. 1800. *Obi; or, The History of Three-Fingered Jack,* ed. Srinivas Aravamudan. Peterborough: Broadview, 2005.

Emerson, Ralph Waldo. 1836. *Nature.* Boston: James Munroe. Internet Archive.

Evernden, Neil. 1992. *The Social Creation of Nature.* Baltimore: Johns Hopkins University Press.

Farrier, David. 2014. Toxic Pastoral: Comic Failure and Ironic Nostalgia in Contemporary British Environmental Theatre. *Journal of Ecocriticism* 6 (2): 1–15.

Garrard, Greg. 1996. Radical Pastoral? *Studies in Romanticism* 35 (3): 449–465.

———. 2004. *Ecocriticism.* London: Routledge.

Gifford, Terry. 1999. *Pastoral.* London: Routledge.

Gilbert, Sandra, and Susan Gubar. 1979. *The Madwoman in the Attic: The Woman Writer and the Nineteenth-Century Literary Imagination.* New Haven: Yale University Press.

Godwin, William. 1794. *Things as They Are; or The Adventures of Caleb Williams*. London: B. Crosby. Internet Archive.

Hawken, Paul. 2007. *Blessed Unrest: How the Largest Movement in the World Came into Being and Why No One Saw It Coming*. New York: Viking.

Hawthorne, Nathaniel. 1850. *The Scarlet Letter*. Boston: Ticknor, Reed, and Fields. *HathiTrust*.

Hiltner, Ken. 2011. *What Else Is Pastoral? Renaissance Literature and the Environment*. Ithaca: Cornell University Press.

Hutchings, Kevin. 2009. *Romantic Ecologies and Colonial Cultures in the British-Atlantic World, 1770–1850*. Montreal: McGill-Queen's University Press.

Lake District National Park. 2019. History of the National Park. http://www. lakedistrict.gov.uk/aboutus/nat_parks_history

Lee, Debbie. 2002. *Slavery and the Romantic Imagination*. Philadelphia: University of Pennsylvania Press.

Leopold, Aldo. 1966. *A Sand County Almanac with Other Essays on Conservation from Round River*. New York: Oxford University Press.

Levine, Caroline. 2015. *Forms: Whole, Rhythm, Hierarchy, Network*. Princeton: Princeton University Press.

McKeon, Richard. 1973. Creativity and the Commonplace. *Philosophy and Rhetoric* 6 (4): 199–210.

McKusick, James. 2000. *Green Writing: Romanticism and Ecology*. New York: St. Martin's Press.

Miller, Carolyn. 2008. The Aristotelian *Topos*: Hunting for Novelty. In *Rereading Aristotle's Rhetoric*, ed. Alan G. Gross and Arthur E. Walzer. Carbondale: Southern Illinois University Press.

Moore, Kathleen Dean, and Scott Slovic. 2014. A Call to Writers. *ISLE* 21 (1): 5–8.

Murphy, Patrick D. 1995. *Literature, Nature, and Other: Ecofeminist Critiques*. Albany: State University of New York Press.

———. 2000. *Farther Afield in the Study of Nature-Oriented Literature*. Charlottesville: University Press of Virginia.

———. 2009. *Ecocritical Explorations in Literary and Cultural Studies: Fences, Boundaries, and Fields*. Lanham: Lexington.

———. 2017. *The Media Commons: Globalization and Environmental Discourses*. Urbana: University of Illinois Press.

Rhodes, Edward Lao. 2003. *Environmental Justice in America: A New Paradigm*. Bloomington: University of Indiana Press.

Ruffin, Kimberly N. 2010. *Black on Earth: African American Ecoliterary Traditions*. Athens: University of Georgia Press.

Shelley, Percy Bysshe. 1820. *Prometheus Unbound, A Lyrical Drama in Four Acts, With Other Poems*. London: C. and J. Ollier. *HathiTrust*.

Shelley, Mary. 1831. *Frankenstein; or, The Modern Prometheus*. London: Colburn and Bentley. Internet Archive.

Southwest Organizing Project. 1990. Letter to Big Ten Environmental Group. *EJNet* March 16.

Sullivan, Heather. 2016. The Dark Pastoral: Goethe and Atwood. *Green Letters: Studies in Ecocriticism* 20 (1): 47–59.

Taylor, Dorceta. 2002. *Race, Class, Gender, and American Environmentalism.* Portland: U.S. Department of Agriculture, Forest Service, Pacific Northwest Research Station.

———. 2014. The State of Diversity in Environmental Organizations. *Green 2.0* July. https://www.diversegreen.org/the-challenge/

Thoreau, Henry David. 2001. *Collected Essays and Poems*, ed. Elizabeth Hall Witherell. New York: Literary Classics of the United States.

West, Thomas. 1780. *A Guide to the Lakes.* 2nd ed. London: Richardson and Urquhart. Internet Archive.

Whitman, Walt. 1855. *Leaves of Grass.* Brooklyn: The Walt Whitman Archive.

Whittier, John Greenleaf. 1846. *Voices of Freedom.* Philadelphia: Thomas S. Cavender. *HathiTrust.*

Wordsworth, William. 1835. *A Guide Through the District of the Lakes in the North of England with a Description of the Scenery, &c. for the Use of Tourists and Residents.* 5th ed. Kendal: Hudson and Nicholson. *HathiTrust.*

CHAPTER 2

Black Nature

HIKING WHILE BLACK

Compared to the U.S. population as a whole, people of color are dispro-
portionately exposed to environmental hazards, such as chemical and particulate air pollutants, unsafe drinking water, poor-quality food, and toxic
neighborhoods. At the same time, there are social and economic barriers
that prevent people of color from accessing environmental benefits like
clean air, water, food, homes, and neighborhoods.[1] Similar barriers block
access to benefits like outdoor recreation to maintain physical well-being
and mental health, as well as a sense of spiritual connection to a natural community. In some cases, the obstacles responsible for environmental injustice
are political and economic, such as when corporations construct toxic waste
sites near impoverished and disenfranchised communities of color. In other
cases, the barriers are conceptual or emotional. People of color are sharply
underrepresented among the nearly 300 million annual visitors to the more
than 400 units of the National Parks System. A 2009 study conducted by
social scientists at the University of Wyoming found that about half of white,
Asian, and Native American respondents reported visiting a national park in
the preceding two years, while less than a third of African American and
Hispanic respondents had done so. The survey also found that, when asked
why they did not visit, African Americans and Hispanics were much more
likely to state not only that the parks are unsafe and uncomfortable, but also
that they "just don't know that much about" them.[2] In other words, the

© The Author(s) 2019
L. Newman, *The Literary Heritage of the Environmental Justice
Movement*, Literatures, Cultures, and the Environment,
https://doi.org/10.1007/978-3-030-14572-9_2

most significant barriers to access for people of color were that they do not feel welcome and that they lack information that would support their engagement with the parks.[3]

Segregation of the national parks, which are the common property of the American people, is not only unjust; it is also a serious obstacle to their preservation. During the Jim Crow era, many national parks were formally segregated, especially in the South. Now, the National Park Service is actively working to increase the diversity of its visitors, in part because of worries about declining support for the agency's mission as the United States makes the demographic transition to being a "minority majority" nation. According to Cliff Spencer, the African American superintendent of Mesa Verde National Park, "we can't allow millions of people, generations of people, to not experience parks and to have no connection to them. When those people get into positions where they'll influence policy and hold the purse strings, they won't understand what parks are and how important they are."[4] As political and economic power become increasingly distributed across racial and ethnic lines in the United States, the Park Service recognizes that it must find a way to enlist diverse bases of support.

One strategy has been to create new units that are focused on the history of people of color, such as the César Chavez National Monument and the Harriet Tubman Underground Railroad National Monument, created in 2012 and 2013 respectively. However, when it comes to already existing parks, the University of Wyoming report recommended that the Park Service "translate increased awareness into greater use [by offering] interpretive programming that relates NPS units to the cultural experiences and interests of specific race/ethnic populations."[5] *A Call to Action*, the working mission statement and strategic plan for the agency's second century, makes a commitment to "welcome and engage diverse communities through culturally relevant park stories and experiences that are accessible to all."[6] In other words, the Park Service has committed itself to environmental humanities programming as a way of taking down the conceptual and emotional barriers that prevent people of color from visiting. Yosemite Interpretive Ranger Shelton Johnson has led the way with his programs about the role of Buffalo Soldiers in the early history of the California national parks. When the Park Service announced that Johnson had received a national award for excellence in interpretation, the press release stated that "by telling the previously untold stories of diverse peoples," he has "facilitated lasting connections between African Americans and their national parks."[7]

There is some measure of simple bureaucratic self-preservation at work here. However, there is much more at stake in the effort to engage communities of color than the continued survival of the Park Service and the units it manages. As *A Call to Action* states, the "National Park System inspires conservation and historic preservation at all levels of American society, creating a collective expression of who we are as a people and where our values were forged."[8] Indeed, for the last century, national parks have served as civic cathedrals where American national identity has been formed during ritualistic visits. An increasingly central feature of that collective identity has been a commitment to protect nature. Visiting and learning about the parks has engaged millions of visitors in the mission of conservation. And the habit of caring about individual sites has translated into commitments to conservation on regional, national, and even global scales. In other words, the National Park Service has provided basic environmental education, fostering public attitudes that have supported the work of the broad environmental movement. In order for the parks to continue playing that role *and* for the movement to continue to thrive, the work of raising awareness and engagement among people of color must continue and must succeed.[9]

Just as importantly, public land management agencies are subject to binding obligations to protect the civil rights of all citizens and must take affirmative action to integrate the places they manage. This is no trivial task, for the truth is that segregation of public lands has much deeper causes than a failure to tell the right stories. In fact, the wilderness has been segregated in the national imagination since it was invented. Paul Outka has demonstrated that, for more than two centuries in the United States, white male identity has been formed through the ritual excursions to the "white-only space" of sublime wilderness. The national parks have served for much of their history as places where whites performed their appreciation of natural beauty as a way of asserting their racial identity.[10] Moreover, the idea of wilderness has always been "saturated with the authority of slavery and the possibility of violent punishment."[11] As a result, African Americans have struggled "to name, frame, and claim a green space" even while they have "actively sought healing, kinship, resources, escape, refuge, and salvation in the land."[12] However, African American historical experience includes not only trauma, but also triumph. In his pioneering study, *Ride Out the Wilderness*, Melvin Dixon writes:

> During slavery blacks depicted the wilderness as a place of refuge beyond the restricted world of the plantation .... The woods or the swamps were regular sites for religious meetings and conversion experiences in which slaves attained important levels of spiritual mobility. Once this covert model for spiritual elevation, release, and self-esteem was established, slaves were more readily able to envision and often secure physical mobility. Escaping North to free territory or overt resistance to slavery usually started with flight deep into the woods.[13]

In addition to preserving the memory of oppression, collective memory can pass down the inspiring legacy of victorious struggle. Given the richness and complexity of African American historical experience, public lands—the commons—must be recognized as fundamentally important cultural resources. Desegregating these heritage landscapes will afford communities of color the civil right to reconnect with symbolic spaces where they forged their collective histories and identities.[14]

Recent interpretive and outreach efforts by the Park Service (and other land management agencies) are welcome signs of change. If these campaigns succeed, many people of color will enjoy restored access to one of the most important environmental benefits, a sense of personal connection to nature. That connection may in turn inspire commitment to protect the places where it is formed. In other words, wilderness is more likely to survive as healthy habitat if it is integrated as social symbolic space. However, if the work of integrating the commons leads to long-term connections between civil rights movements and environmental movements, the potential benefits will be far greater than preservation of habitat and biodiversity. As James Gustav Speth and J. Phillip Thompson put it in the title of a recent article, "A Radical Alliance of Black and Green Could Save the World."[15] After all, only activist communities that can build broad solidarity across difference will be able to confront globalization and global warming. In "Why #BlackLivesMatter Should Transform the Climate Debate," Naomi Klein writes, "Racism is what has made it possible to systematically look away from the climate threat for more than two decades." Only multiracial movements that explicitly confront racism and advocate for environmental justice can "jolt us out of our climate inaction."[16]

Fortunately, the work of integrating nature and connecting movements is well underway. In 1991, the First National People of Color Environmental Leadership Summit promulgated the influential manifesto, "Principles of Environmental Justice."[17] Since then, more and more activists across the United States and beyond have stepped forward to reclaim a place on the land for people of color and to help reorient the mainstream environmental movement toward their concerns. For instance, Angelou Ezeilo founded the Greening Youth Foundation "to work with diverse, underserved and underrepresented children, youth and young adults in an effort to develop and nurture enthusiastic and responsible environmental stewards." The organization delivers environmental education programming and partners "with land management agencies to provide service and internship opportunities for youth and young adults thereby creating pathways to conservation careers."[18] Similarly, Roberto Moreno has created the Camp Moreno Project "for the purpose of diversifying the out of doors, attracting and engaging urban youth [and] encouraging park use by underserved audiences."[19] Camp Moreno has organized "Camping 101" get-togethers at Rocky Mountain, Grand Canyon, Saguaro, and other national parks, and these events have given tens of thousands of young people of color their first overnight outdoor experiences. Likewise, Rue Mapp has organized *Outdoor Afro*, a thriving online social network "that reconnects African-Americans with natural spaces and one another through recreational activities such as camping, hiking, biking, birding, fishing, gardening, skiing."[20] Other initiatives include the Diverse Environmental Leaders National Speakers Bureau, which represents "a wide range of talented and accomplished environmental professionals of color who can help shift the environmental conversation in America to become more inclusive and equitable at all levels."[21] Hundreds of organizations like these across the United States carry on the struggle for integration of the commons, at the same time that they are creating the next generation of environmental activists and organizers of color.[22]

There are important ways that scholars in the environmental humanities can support the work of environmental integration. A decade ago, Lawrence Buell called for an "ecojustice revisionism" in environmental literary studies, and that work is well underway.[23] There is no shortage of powerful stories for environmental humanists to recover and retell.[24] Camille Dungy's anthology *Black Nature: Four Centuries of African American Nature Poetry*—the first book of its kind—demonstrates the power and richness of this long-lived and complex literary tradition.[25] Dungy observes that the archive she curates is characterized by "poems written from the perspective of the workers of the field [that] describe moss, rivers, trees, dirt, caves, dogs, fields: elements of an environment steeped in a legacy of violence, forced labor, torture, and death."[26] While the poems in *Black Nature* maintain a critical distance from Nature, they also record close observation, intimate familiarity, and deep appreciation for the land. In *Black on Earth*, Kimberly Ruffin notes the same tension and calls it the "ecological beauty-and-burden paradox":

> People of African descent endure the burden and enjoy the beauty of being natural. They bear the burden of a historical and present era of environmental alienation [and they] struggle against the burden of societal scripts that make them ecological pariahs, yet they enjoy the beauty of liberating themselves and acting outside of these scripts. Their ecological outlook is informed both by the collective experience of being placed among those at the bottom of human hierarchies and their visionary responses to nature itself.[27]

Learning to understand those experiences *and* responses will help environmentalists meet the needs of the present moment. For instance, African American environmental literature includes a strong emphasis on work, as opposed to leisure or recreation; learning to appreciate labor as a way of interacting with nature would deepen contemporary environmentalism, allowing it to connect with the experience and concerns of many people who are currently alienated by what they perceive as the movement's elitism.[28] In other words, as Ruffin writes, learning to recognize "cultural variants of key concepts such as nature [and] acknowledging racial/ethnic diversity within U.S. environmental history can better prepare us to interact in national and transnational conversations about the environment."[29] There is no better place to begin that work than in the writings of the black abolitionists of the nineteenth century, who showed that slavery was not only racist genocide, but also an ecological disaster.[30]

Anti-slavery Gothic, Radical Pastoral, and Revolutionary Sublime in Slave Narratives

In *Slavery in the United States; A Narrative of the Life and Adventures of Charles Ball, a Black Man* (1837), Ball is sold onto a cotton plantation in the Deep South. Soon after he arrives there, two other slaves, David and Hardy, kidnap the owner's daughter and hide her in a nearby swamp. A posse captures the two men and, after debating the most excruciating way to punish them, they stake their captives to the ground and leave them to die. The woman is discovered nearby, and she dies a few days later from causes that Ball leaves unstated. For the remainder of the narrative, "Murderer's Swamp" is haunted by the memory of the abductors' fate and by their skeletons, which have been picked clean by carrion crows. Some months later on a Sunday afternoon, Ball is permitted to go hunting for game to supplement his rations. He hears the mysterious sound of bells in the swamp, and a "strange spectre" appears:

> I saw come from behind a large tree, the form of a brawny, famished-looking black man, entirely naked, with his hair matted and shaggy, his eyes wild and rolling, and bearing over his head something in the form of an arch, elevated three feet above his hair, beneath the top of which were suspended the bells, three in number, whose sound had first attracted my attention.[31]

Ball's swelling and rhythmic description allows the reader to share his terror at this apparition, before he reveals that the man is "a poor destitute African negro" named Paul whose back is "seamed and ridged with scars of the whip."[32] After multiple escapes, Paul has been forced to wear an iron collar with a hoop of bells that encircles his head. Nevertheless, he has escaped again and is hiding deep in the woods. As the truth about him emerges, his presence, as an emblem of the slave-owners' deliberate malice and cruelty, reinforces the gothic terrors of the Southern woods. Ball gives Paul several terrapins he has caught, lights a fire so he can roast them, and promises to return. As he makes his way back to the plantation, Ball describes the scene: "the night-breeze agitated the leaves of the wood and moaned in dreary sighs through the lofty pine tops; the gale shook the forest in the depth of its solitudes: a cloud swept across the moon, and her light disappeared; a flock of carrion crows disturbed in their roosts, flapped their wings and fluttered over my head; and a wolf ... greeted the darkness with a long and dismal howl."[33] Not just cruel owners, but hostile wild animals terrorize the slaves of the South, threatening to kill and consume the bodies of slaves who attempt to escape and take refuge in nature. On the following Sunday, Ball returns to the swamp with a file, hoping to help Paul remove his collar. There, he sees "a large sassafras tree, around the top of which was congregated ... an assembly of the obscene fowls of the air." In the tree, the "lifeless and putrid body of the unhappy Paul hung suspended by a cord made of twisted hickory bark, passed in the form of a halter round the neck, and firmly bound to a limb of the tree."[34] His flesh has remained on his bones long enough to rot because the bells scare off any birds that attempt to alight on his body, which remains hanging from the tree, permanently haunting the dismal swamp in which he sought freedom.

The anti-slavery gothic is an imaginative landscape saturated with blood and littered with bones. As a literary *topos*, it forces the readers of slave narratives and other abolitionist texts to acknowledge and even vicariously experience the terrors of Southern slavery. However, it also has the unavoidable effect of emphasizing the vulnerability and helplessness of the slaves, who are surrounded and immobilized by a hateful nature that actively colludes with the slave-owners. Not surprisingly then, black writers in the Romantic era experimented with other ways of imagining and writing about nature. For instance, in the Methodist conversion narrative, *A Narrative of the Lord's Wonderful Dealings with John Marrant, A Black* (1785), the wilderness on the margins of the colonial South serves the young Marrant as a refuge of liberty where he can practice his enthusiastic faith without being ridiculed by his family and friends. He discovers this feeling of freedom on a journey from Charleston to rural Georgia, when he experiences "much communication with God on the road."[35] So when his family's harassment becomes especially intense, he retreats to "the fields, and some days staid out from morning to night to avoid the persecutors ... but seemed to have clearer views into the spiritual things of God."[36] After deciding that "it was better for me die than to live among such people," Marrant "crosses the fence, which mark[s] the boundary between the wilderness and the cultivated country."[37] He spends several months in the forests, where he survives by eating grass and sleeping in trees to avoid wolves and bears. A native hunter takes the lost youth under his wing, teaches him the Cherokee language and basic survival skills, and then brings him to "a large Indian town."[38] There, Marrant is condemned to death, but saves himself by preaching in Cherokee, converting his executioner, and healing the "king's daughter" through prayer. Marrant is adopted into the tribe as a spiritual leader: "I had assumed the habit of the country, and was dressed much like the king, and nothing was too good for me."[39] He tests his new status by making a two-month missionary journey to three tribes who live west of the Cherokee. Finally, he decides to head home to Charleston. On his return journey through the wilderness, he is a confident and self-reliant woodsman. His transformation is so complete that when he arrives, his family and friends fail to recognize him: "My dress was purely in the Indian stile; the skins of wild beasts composed my garments, my head was set out in the savage manner, with a long pendant down my back, a sash round my middle without breeches, and a tomahawk by my side." He gathers a congregation in "the back settlements of the white people," where his hybrid cultural identity gives him unique influence.[40] He manages to convert one back-slider simply by scolding him for failing to say grace: "Here is a wild man, says he, come out of the woods to be a witness for God, and to reprove our ingratitude and stupefaction!"[41] Marrant's narrative was published in London to drum up financial support for his mission to the Mi'kmaq in Nova Scotia. And throughout his tale, the forests of the South figure as the ground of his conversion from a timid neophyte into a powerful instrument of God's will who will ensure "that vast multitudes of hard tongues, and of a strange speech, may learn the language of Canaan, and sing the song of Moses, and of the Lamb."[42] The book does not explicitly address the political issues of slavery and the displacement of Native Americans. However, it does have important anti-racist implications, since it represents Marrant as a religious leader of color whose authority is rooted in his experience of divinity in nature. Throughout his autobiography, Marrant describes the forests of the South as revolutionary sublime refuges where escaped slaves and natives live free of harassment by the whites who have colonized the seaboard.

Another vision of black independence and self-possession in nature comes in *The Interesting Narrative of the Life of Olaudah Equiano, or Gustavus Vassa, the African*. Published in 1789, this autobiography begins with an ethnographic sketch of the "manners and customs" of the Igbo people in the village of Essaka in what is now Nigeria.[43] Equiano's primary rhetorical goal in this opening chapter is to establish his authority by refuting his European readers' racist assumptions about Africans. Thus he adopts the voice of an Enlightenment natural historian and describes Igbo ways of life in terms that explicitly invite comparison with European customs: "The dress of both sexes ... generally consists of a long piece of callico, or muslin, wrapped loosely round the body, somewhat in the form of a highland plaid."[44] By comparing the Igbo to the Gaelic tribes of the Scottish highlands, Equiano categorizes them according to Romantic ideas of civilization and primitivism. They are the wild but noble inhabitants of a definitively natural landscape. Equiano varies this theme by harnessing the radical pastoral *topos* that was circulating in London literary and political circles at the time. He describes Essaka as "a charming fruitful vale" where the land is abundantly productive:

> We have plenty of Indian corn, and vast quantities of cotton and tobacco. Our pine apples grow without culture; they are about the size of the largest sugar-loaf, and finely flavoured. We have also spices of different kinds, particularly pepper; and a variety of delicious fruits which I have never seen in Europe; together with gums of various kinds, and honey in abundance.[45]

The fertility that Equiano nostalgically catalogs allows the people of Essaka to live a virtuous life: "our manners are simple, our luxuries are few." They work hard, exerting their "industry ... to improve [the] blessings of nature."[46] And they are free of urban luxury and vice: "As we live in a country where nature is prodigal of her favours, our wants are few and easily supplied." The Igbo are farmers, and social and economic relations in Essaka are communal and egalitarian: "we are all habituated to labour from our earliest years [and everyone] contributes something to the common stock."[47] In this natural republic, "every transaction of the government ... was conducted by the chiefs or elders of the place," who also "decided disputes and punished crimes; for which purpose they always assembled together." Equiano goes on to catalog Igbo religious beliefs and cultural traditions, in terms carefully chosen to simultaneously exoticize and familiarize this "nation of dancers, musicians, and poets."[48]

All in all, the first chapter of the *Interesting Narrative* presents the Igbo as a free and independent people thriving in a healthy homeland that they cultivate carefully. This radical pastoral vision then serves as a benchmark against which he implicitly measures the slave-based society of the southern colonies as he narrates his extensive travels. Equiano identifies himself as "the African" in the book's title, and throughout he observes and comments on Southern cruelty and greed from the rational and sympathetic perspective of a virtuous man of the soil. Thus, when the slaves with whom he has made the Middle Passage are auctioned off in Barbados, Equiano first laments the practice of separating families and then reasons with the reader about it: "Why are parents to lose their children, brothers their sisters, or husbands their wives? Surely this is a new refinement in cruelty, which, while it has no advantage to atone for it, thus aggravates distress, and adds fresh horrors even to the wretchedness of slavery." Equiano uses the same combination of warm compassion and cool analysis when he describes overwork, poor housing, starvation, rape, and torture in agonizing detail and then deplores them as "unmerciful, unjust, and unwise."[49]

Slave narratives, like those written by Ball, Marrant, and Equiano, sold many thousands of copies in the first half of the nineteenth century, especially at conferences, fairs, and other public events organized by antislavery activists. Together they form one of the most disturbing and stirring archives of Transatlantic Romanticism. In his autobiographies and other writings, Frederick Douglass drew on this archive for inspiration, adapting among other elements the rich variety of landscape *topoi* that he found there.

FREDERICK DOUGLASS, NATURE, AND ABOLITION

If the overthrow of slavery in the United States was the most revolution-ary event of the nineteenth century, then Frederick Douglass, the period's leading black abolitionist, was its most important literary and historical figure.[50] As sectional tensions flared during the late 1840s and especially after the Compromise of 1850, free blacks in the North faced intensifying hatred and violence. In response, Douglass hoped to broaden the aboli-tionist movement by building multiracial solidarity in opposition to slav-ery. He did so by repurposing the radical pastoral and the revolutionary sublime in order to appeal to participants in the broad Free Soil move-ment.[51] He also adopted the core of Free Soil doctrine, which centered on the belief that liberty achieved its truest expression when free people could mix their labor with nature in the pursuit of self-reliance. Therefore, dem-ocratic access to arable land was a precondition of genuine political and economic emancipation. Capitalism had alienated land that had once been held as commons and had concentrated it into ever larger parcels of pri-vate property. Abolishing the exploitation of labor, both slave and wage, would require reversing this process.

Douglass set out to build a wide following for his ideas about labor and nature by fictionalizing them in his second autobiography, *My Bondage and My Freedom*, and in his only novella, *The Heroic Slave* (1853), the fictional-ized story of Madison Washington, leader of the 1841 *Creole* mutiny. The rhetorical functions of the landscapes in these texts from the mid-1850s can be seen especially clearly in comparison with his first autobiography, *The Narrative of the Life of Frederick Douglass, an American Slave* (1845). In the *Narrative*, nature is an anti-slavery gothic wilderness, while in the later novella and autobiography, the forests of Virginia are rendered as revolu-tionary sublime theaters of self-emancipation.[52] The differences between the earlier and later texts correspond closely with Douglass's political evo-lution. He began his career believing that changing minds one by one would gradually transform broad social attitudes about slavery and lead to emancipation. But he later came to see that large-scale organized political pressure would be necessary, so he set out to nurture existing anti-slavery commitments within the most vibrant working-class political milieu of the 1840s and 1850s, the Free Soil movement.[53] Douglass's imaginative land-scapes and powerful ideas launched the black agrarian political tradition and offer continuing insight into the distinctive environmental concerns of people of color. His integration of the revolutionary sublime and the radical pastoral *topoi* into abolitionist discourse is not only an important example of an alternative black tradition; it also exemplifies the organic connection between the struggles for social and environmental justice in the antebellum United States and beyond.

A year after Douglass's 1838 escape from slavery, he joined William Lloyd Garrison's Massachusetts Anti-Slavery Society, an organization dedicated to ending slavery by changing people minds through moral "suasion," as opposed to political action. In the *Narrative*, Douglass describes his feelings of timidity when telling his story in front of his first audience: "the idea of speaking to white people weighed me down." After a time, though, he "felt a degree of freedom, and said what [he] desired with considerable ease."[54] The phrase "a degree of freedom" indexes his position as a working-class black speaker employed by Garrison's elite white organization. James McCune Smith describes Douglass's subordination in this new context, writing that he was "a glorious waif to those most ardent reformers."[55] Even among those reputed to be the most radical of American abolitionists, he faced intense prejudice: "these gentlemen, although proud of Frederick Douglass, failed to fathom, and bring out to the light of day, the highest qualities of his mind; the force of their own education stood in their own way: they did not delve into the mind of a colored man for capacities which the pride of race led them to believe to be restricted to their own Saxon blood."[56] Garrison pressured Douglass to tell audiences the story of his life in conformity with a rigid doctrine of nonresistance that rejected political action in favor of moral suasion and, according to Smith, Douglass became, "after the strictest sect, a Garrisonian."[57] As a result, the *Narrative* toes a nonresistant line, and Douglass consistently casts himself as a martyr, a passive victim whose salvation depends on slaveholders capitulating to the moral indignation of his Northern readers.

Central to his self-construction in this role is the anti-slavery gothic *topos* that he deploys to emphasize his isolation, vulnerability, and misery. The "deep pine woods" of Talbot County echo with the "wild songs" slaves sing to relieve their sorrows. To Douglass, this forlorn music sounds like the "singing of a man cast away upon a desolate island."[58] In these woods where "all is gloom," his grandmother is turned out to die in "a little hut" with "a little mud chimney." Instead "of the voices of her children, she hears by day the moans of the dove, and by night the screams of the hideous owl."[59] The pine woods form a purgatorial borderland that surrounds Douglass's known world and serves not only as a wall between

North and South but also as the definitive place of exile. He writes that when, as a young man, he allowed himself to "survey the road" to freedom, he imagined a howling wilderness of the worst sort:

> Upon either side we saw grim death, assuming the most horrid shapes. Now it was starvation, causing us to eat our own flesh; – now we were contending with the waves, and were drowned; – now we were overtaken, and torn to pieces by the fangs of the terrible bloodhound. We were stung by scorpions, chased by wild beasts, bitten by snakes, and finally, after having nearly reached the desired spot, – after swimming rivers, encountering wild beasts, sleeping in the woods, suffering hunger and nakedness, – we were overtaken by our pursuers, and, in our resistance, we were shot dead upon the spot![60]

The breathless syntax of this catalog of horrors puts the reader in the shoes of an escaped slave, racing through an anarchic landscape where violent death lurks, like Edward Covey the slave-breaker, "under every tree, behind every stump, in every bush."[61] And rather than arrive finally at freedom at the end of the passage, the reader collides with a monosyllabic finality, death. This is anti-slavery gothic at its most powerful.

Even the climactic tenth chapter of the *Narrative*, during which Douglass defends himself against Covey, takes place in this gothic forest. Douglass represents himself, newly arrived from Baltimore by way of Thomas Auld's plantation, as an urban bumpkin, "even more awkward than a country boy appeared to be in a large city." By way of illustrating his awkwardness, Douglass opens the chapter with an anecdote about losing control of Covey's ox cart in the woods. Pausing to note the pathos of the tableau, he remarks, "There I was, entirely alone, in a thick wood, in a place new to me. My cart was upset and shattered, my oxen were entangled among the young trees, and there was none to help me."[62] To make matters worse, this lonely and disorienting landscape is ruled by a satanic reptile, Covey, who lies "coiled up in the corner of the wood-fence" watching his slaves work and looking for opportunities to whip them.[63] When Douglass runs away from Covey for the first time, his path leads him through alliterative "bogs and briers, barefooted and bare-headed." When he arrives at his owner's plantation, he lingers on a sketch of his wretched state: "My hair was all clotted with dust and blood; my shirt was stiff with blood. My legs and feet were torn in sundry places with briers and thorns, and were also covered with blood. I suppose I looked like a man who had escaped a den of wild beasts." Finally, after recognizing that he cannot "stay in the woods" or he will "be starved to death," Douglass decides to "go home" to Covey's farm, where he expects to be "whipped to death."[64] Throughout the *Narrative*, Douglass represents the Southern woods as a desolate, life-threatening wilderness in which slaves are powerless to defend themselves. By implication they require the assistance of white Northern abolitionists.

There are important moments when the *Narrative* cuts against its overall emphasis on righteous passivity. Above all, the fight scene with Covey foregrounds an act of physical resistance to slavery. Even here though, Douglass carefully contains slave violence within the bounds of self-defense, staging the scene in such a way that he never takes the initiative. Rather than decisively thrash his opponent, Douglass fights him to a draw, doing no more than parry Covey's blows. In the end, this fight scene allegorizes the overthrow of slavery by Garrisonian means: Douglass wins Covey's respect by mounting a righteous defense against unfair treatment. As a result he is treated more humanely for the remainder of his time on Covey's farm.[65] Similarly, the anti-slavery gothic character of the *Narrative* is undercut by the implication that slaves' fears of nature are no more than psychological shackles. By melodramatizing the terrors of the road North (note, for instance, the rhythmic repetition of the words "blood" and "wild beasts" in the passages quoted above), Douglass emphasizes the naiveté of his youthful self, and he suggests that the image of the howling wilderness is one more tool used by the masters to imprison slaves, who must, as a necessary condition of self-emancipation, become aware of their own *potential* independence in the world. He follows through on this modest suggestion in his novella, *The Heroic Slave*, which was published eight years after the *Narrative* in 1853.

THE HEROIC SLAVE

Despite the dominance of Garrisonian doctrines of nonresistance, some abolitionists, both black and white, cheered Nat Turner's 1831 insurrection, Joseph Cinque's 1839 uprising on the *Amistad*, and Madison Washington's 1841 mutiny on the *Creole*.[66] By way of engaging this emergent militancy, Douglass repurposes the discourse of republicanism in *The Heroic Slave*, where he presents a fictionalized Madison Washington as a revolutionary hero fighting in a latter-day war of independence.[67] Washington escapes from slavery in Virginia and makes his way to Canada with the help of a white Ohioan named Listwell. Unable to bear the thought of his wife's continued bondage, he returns to the South to rescue her. He is captured in the attempt and, while being transported to New Orleans, he leads a mutiny, taking control of the slave ship *Creole* and sailing it to freedom in the British Bahamas.[68]

Douglass casts Madison Washington as a figure of sublime obscurity, a latent popular hero who has been unjustly expunged from national memory: "Glimpses of this great character are all that can now be presented. He is brought to view only by a few transient incidents, and these afford but partial satisfaction." Washington shines as a beacon of liberation for his oppressed people, for "like a guiding star on a stormy night, he is seen through the parted clouds and the howling tempests." However, he is more than a passive light to steer by, for "like the gray peak of a menacing rock on a perilous coast, he is seen by the quivering flash of angry lightning, and he again disappears covered with mystery."[69] Douglass positions Washington, the rebel slave, as the single stable object in a landscape of sublime mutability. This is a far cry from the anti-slavery gothic of the *Narrative*. Instead, in *The Heroic Slave*, Douglass relies on the sublime to establish the revolutionary character of the novel's hero. Nature is a sacred space where Madison Washington experiences an epiphany that produces militant political consciousness. He first articulates his desire for freedom in a clearing in a "dark pine forest" near "a sparkling brook." It is a Sunday morning and, while his master attends church, Washington seeks relief "among the tall pines" from the pain and humiliation of a whipping. He is secretly observed by Mr. Listwell, who listens to a soliloquy that he delivers after meditating on the freedom of small forest animals: "Those birds, perched on yon swinging boughs, in friendly conclave, sounding forth their merry notes in seeming worship of the rising sun, though liable to the sportsman's fowling-piece, are still my superiors. They live free, though they may die slaves. They fly where they list by day, and retire in freedom at night." The freedom of the birds goads Washington into an enraged awareness of the raw irony of human enslavement. A few sentences later, his attention turns to a snake: "How mean a thing am I. That accursed and crawling snake, that miserable reptile, that has just glided into its slimy home, is freer and better off than I. He escaped my blow, and is safe. But here am I, a man, – yes, a man! – with thoughts and wishes, with powers and faculties as far as angel's flight above that hated reptile, – yet he is my superior."[70] Washington responds to this painful contrast by making an impassioned vow: "*Liberty* I will have, or die in the attempt to gain it …. I have nothing to lose. If I am caught, I shall only be a slave.

If I am shot, I shall only lose a life which is a burden and a curse. If I get clear, (as something tells me I shall,) liberty, the inalienable birth-right of every man, precious and priceless, will be mine. My resolution is fixed. *I shall be free.*[71] Listwell continues to watch as Washington delivers "denunciations of the cruelty and injustice of slavery; heart-touching narrations of his own personal suffering, intermingled with prayers to the God of the oppressed for help and deliverance."[72] This sermon, delivered in a "solitary temple," converts Listwell to abolitionism, and he swears to atone for his "past indifference to this ill-starred race, by making such exertions as I shall be able to do, for the speedy emancipation of every slave in the land."[73] In *The Heroic Slave*, then, the "wildering woods" of Virginia serve as a chapel of liberty. In order to achieve the determination and courage he needs to emancipate himself, Washington "shuns the church, the altar, and the great congregation of christian worshippers, and wanders away to the gloomy forest, to utter in the vacant air complaints and griefs, which the religion of his times and his country can neither console nor relieve."[74] For Douglass, who bitterly condemned pro-slavery churches in the *Narrative*, nature has become a counter-image, a sacred space of authentic spiritual experience and liberty where his protagonist awakens to class consciousness and revolutionary commitment.

The forest also allows Washington to discover his own capacity for self-reliance. When he first escapes from his master's plantation, he does not find the forest welcoming. During his first week of freedom, he is disoriented and miserable. Clouds obscure the North Star, and he wanders in circles, returning eventually to the place from which he had escaped. He almost loses his nerve during this "severe trial" in the wilderness:

> I arrived at home in great destitution; my feet were sore, and in travelling in the dark, I had dashed my foot against a stump, and started a nail, and lamed myself. I was wet and cold; one week had exhausted all my stores; and when I landed on my master's plantation, with all my work to do over again, – hungry, tired, lame, and bewildered, – I almost cursed the day that I was born.[75]

Despite his misery and weakness, Washington resolves to stay in the wilderness after he peers secretly into the slaves' quarters: "I saw my fellow-slaves seated by a warm fire, merrily passing away the time, as though their hearts knew no sorrow. Although I envied their seeming contentment, all wretched as I was, I despised the cowardly acquiescence in their own degradation which it implied, and felt a kind of pride and glory in my own desperate lot." Unable to bear returning to slavery after being transformed by the experience of freedom, however painful and brief, Washington resolves to make "the woods my home" and learns to survive by relying on his own wits. He lives in the wilderness for five years, hiding in a cave by day and wandering "at night with the wolf and the bear."[76] On the one hand, this experience is one of privation; on the other, "the pine woods" are a space of self-determination where Washington's knowledge of the land allows him to remain free even in the midst of tyranny. He is even able to visit his wife once a week, and he goes so far as to tell Listwell, "I had partly become contented with my mode of life, and had made up my mind to spend my days there."[77]

By taking refuge in the woods, Washington has achieved a subversive kind of freedom within the borders of the slave South. But the story cannot stop here, with its hero permanently in hiding within a society bent on his destruction. So, the deus ex machina of an "awful conflagration" springs suddenly to life: "the wilderness that sheltered me thus long took fire, and refused longer to be my hiding-place." This forest fire warrants a strangely elaborate and impassioned description during which Douglass envisions apocalyptic natural retribution for the national sin of slavery: "The whole world seemed on fire, and it appeared to me that the day of judgment had come; that the burning bowels of the earth had burst forth, and that the end of all things was at hand." Virginia's landscape of oppression is scoured clean, as nature takes upon itself the sins of the slavocracy and then self-immolates. Racing to escape the fire, Washington stops for a moment to savor the destruction of the corrupt South:

I looked back to behold [the fire's] frightful ravages, and to drink in its savage magnificence. It was awful, thrilling, solemn, beyond compare. When aided by the fitful wind, the merciless tempest of fire swept on, sparkling, creaking, cracking, curling, roaring, out-doing in its dreadful splendor a thousand thunderstorms at once. From tree to tree it leaped, swallowing them up in its lurid, baleful glare; and leaving them leafless, limbless, charred, and lifeless behind.[78]

Washington is more than a mere witness to this inferno. Chanting a cadence of sublimity—"awful, thrilling, solemn"—he *participates* vicariously in the symbolic erasure of slavery. Strangely he dwells most intently on how the fire ravages the forest animals, describing how "wild beasts of every name and kind, – huge night-birds, bats, and owls ... perished in that fiery storm [along with the] long-winged buzzard, and croaking raven." It is as though the very creatures who, in the anti-slavery gothic, collaborate with the masters must be destroyed by sacrificial violence that spares only the most pure and innocent. The scene ends with Washington watching as "countless myriads of small birds ... rose up to the skies, and were lost to the sight in clouds of smoke and flame." Like these tokens of freedom, the same ones that had initially inspired him to run away, Washington is finally driven from his wilderness retreat by the cataclysm. After seeking "among wild beasts the mercy denied by [his] fellow men," after seeing "his dwelling-place and city of refuge reduced to ashes forever," he runs "alike from fire and from slavery" to freedom in the North.[79] The revolutionary sublime has done its work. He has been transformed from a submissive victim into a redeemer, a hero who, we now feel sure, will defeat slavery with decisive action.[80]

At the end of *The Heroic Slave*, Douglass hints at the unstable character of the Romantic *topoi* he employs. After Washington is burned out of the woods, he makes his way to Ohio, where he meets Listwell again. He relates an anecdote about settling down one night to sleep in a tree when he hears people approaching. "Upon my word," he remarks, "I dreaded more these human voices than I should have done those of wild beasts." From his vantage point in the tree, Washington watches two interactions between slaves and nature. First, a group of "men, who were all colored, halted at least a hundred yards from me, and began with their axes, in right good earnest, to attack the trees. The sound of their laughing axes was like the report of as many well-charged pistols. By and by there came down at least a dozen trees with a terrible crash. They leaped upon the fallen trees with an air of victory."[81] Unaware of Washington and oblivious to the epiphany he has achieved, these slaves participate willingly in their own oppression by desecrating the revolutionary sublime forest. Next Washington watches as one of these woodcutters stays behind after the rest leave and, kneeling before a tree, begins to pray: "'O thou,' said he, 'that hearest the raven's cry, take pity on poor me! O deliver me! O deliver me! in mercy, O God, deliver me from the chains and manifold hardships of slavery! With thee, O Father, all things are possible. Thou canst stand and measure the earth,'"[82] This tableau reverses the novel's opening scene in which Listwell watches Washington praying in the woods. Rather than arrive at a self-reliant resolution to free himself, this slave prays passively and ineffectually for divine deliverance. There is nothing automatic then about the way that nature transforms Washington into a republican hero. Perhaps this explains why, if only when Madison finally arrives in Canada, that he feels confident that he "is out of the woods at last."[83]

DOUGLASS AND FREE SOIL

Douglass's differing decisions about how to represent nature parallel key differences in his thinking during two very different periods in his political life. During the profound economic recession that followed the financial collapse of 1837, he was blackballed from his shipyard trade by white caulkers in New Bedford. Despite this personal experience with white working-class racism, Douglass consistently looked for ways to connect abolitionism with the period's increasingly radical workers' movement, especially after he began to pull himself out of the Garrisonian orbit. After all, the wage-earning class during the antebellum period was diverse along lines of both race and gender, and many, though by no means all, workers supported and even fought for the liberation of women and people of color.[84] Douglass came to believe that abolition "is a poor man's work." "It is not to the rich that we are to look," he wrote, "but to the poor, to the hardhanded working men of the country, these are the men who are to come to the rescue of the slave."[85] He formed this attitude in part during his 1845–1847 travels in England, Scotland, and Ireland, when he worked with Henry Vincent and William Lovett, key leaders of the militant working-class movement, Chartism.[86] The title of Douglass's newspaper honors Feargus O'Connor's *Northern Star*, the foremost Chartist periodical, and Douglass repeatedly allies himself with Chartism in articles calling for self-emancipation through mass action.[87] He also describes England as the place where he recognized how relevant the international republican political and philosophical tradition was to the struggle to abolish slavery:

> It so happened that the great mass of the people in England who attended and patronized my anti-slavery meetings, were, in truth, about as good republicans as the mass of Americans, and with this decided advantage over the latter – they are lovers of republicanism for all men, for black men as well as for white men. They are the people who sympathize with Louis Kossuth and Mazzini, and with the oppressed and enslaved, of every color and nation, the world over.[88]

In addition to transforming his political ideas, Douglass's time overseas radically shifted his way of thinking about the landscape of his home country:

> In thinking of America, I sometimes find myself admiring her bright blue sky, her grand old woods, her fertile fields, her beautiful rivers, her mighty lakes, and star-crowned mountains. But my rapture is soon checked, my joy is soon turned to mourning. When I remember that all is cursed with the spirit of slaveholding, robbery, and wrong; when I remember that with the waters of her noblest rivers, the tears of my brethren are borne to the ocean, disregarded and forgotten, and that her most fertile fields drink daily of the warm blood of my outraged sisters; I am filled with unutterable loathing.[89]

Douglass wrote this searing passage in a letter to Garrison, posted from the Victoria Hotel in Belfast on New Years' Day in 1846 and later published in the *Liberator*. He addresses his former mentor as "My Dear Friend" and relates a series of anecdotes that simultaneously perform his newfound independence and condemn the racism of the Northern United States by contrast with Ireland, where he has been welcomed warmly as an equal. In short, Douglass's two-year residence in Britain was a decisive turning point. As he wrote to Garrison, "I seem to have undergone a transformation. I live a new life."[90] He was no longer a subordinate spokesperson for a white-dominated organization; instead, he was an independent black abolitionist leader.[91]

Released from the fetters of sectarianism, Douglass rejected Garrison's opinions on the crucial question of political action.[92] He broadened his commitments and activism, making connections between various forms of oppression and looking for ways to unite constituencies across arbitrary divisions. He was one of the antebellum period's most active male feminists, he advocated temperance, he opposed flogging in the Navy, he called himself a Chartist, and he endorsed the working-class pacifist Elihu Burritt's campaign for universal peace. In short, he immersed himself in the multifaceted transatlantic movement for social change that defined the political climate of the 1840s and that reached a climax in the revolutions of 1848.[93] During this time, he developed a radical interpretation of the Declaration of Independence and the U.S. Constitution as anti-slavery documents. This effort came to fruition in one of his most influential public addresses, the 1852 speech, "What to the Slave Is the Fourth of July?" in which he asks, "Must I argue the wrongfulness of slavery? Is that a question for Republicans?" He answers that, at "a time like this, scorching irony, not convincing argument, is needed. O! had I the ability, and could I reach the nation's ear, I would, to-day, pour out a fiery stream of biting ridicule, blasting reproach, withering sarcasm, and stern rebuke. For it is not light that is needed, but fire."[94] This kind of militant republicanism would remain the dominant strain in Douglass's thinking at least until the conclusion of the Civil War.[95]

Soon after he returned to the United States to establish the *North Star* in Rochester, New York, Douglass came into contact with a sharply different abolitionist milieu than he had known in Massachusetts. Western New York during the first half of the 1840s was one of the strongholds of the Liberty Party, the anti-slavery organization founded by key political abolitionists after the split with Garrison and endorsed by prominent black abolitionists such as Theodore S. Wright, Henry Highland Garnet, Henry Bibb, and Samuel Ringgold Ward.[96] Though it was extremely effective as a kind of long-term propaganda campaign on the question of slavery and managed to place the issue irreversibly on the national agenda, the Liberty Party's single-issue platform sharply limited its electoral appeal.[97] By 1847, when Douglass arrived in Rochester, most political abolitionists had recognized that in order to build a political force with the kind of power that it would take to overthrow slavery, it would be necessary to integrate abolitionism into a fully developed working-class political program.[98] The Free Soil Party was founded in August 1848 in Buffalo, New York at a convention attended by more than 10,000 supporters. There was substantial support from black abolitionists, including Douglass, who was invited to address the assembled crowd.[99] At the convention, Barnburner Democrats joined with Conscience Whigs and Liberty Party members to denounce the collusion of Southern plantation owners with Northern factory owners, whom they lumped together as "the Slave Power." The speakers argued that a new form of tyranny had taken root when a few hundred thousand slaveholders could control the economy and write the laws of a free republic of millions. Moreover, slavery represented an intensified capitalism, since the slaveholders applied the raw logic of the marketplace to the production and distribution not only of commodities, but also of labor. Decisions about human lives were made exclusively on the basis of cold calculations of profit and loss.[100] The Free Soil conventioneers adopted a platform that combined anti-slavery demands with populist republicanism under the slogan "Free Soil, Free Speech, Free Labor and Free Men." The new formed Free Soil Party took its name from the central principle of the Wilmot Proviso, that slavery should be barred from new territory acquired as a result of the invasion of Mexico. Running against two pro-slavery candidates, Zachary Taylor and Lewis Cass, in the election of 1848, Free Soil's nominee, Martin Van Buren, tallied almost 300,000 votes, a full 10% of the total cast. Despite their defeat in this elections, the broad Free Soil movement surrounding the party was the most prominent and dynamic progressive force in the United States for the next eight years.

According to the discourse of Free Soil, liberty reached its pinnacle when a free man mixed his labor with nature in pursuit of economic self-reliance and happiness. Since most Free Soilers also believed that political equality was meaningless without substantive economic equality, they extended the meaning of the phrase "Free Soil," making it also call for the federal government to guarantee the natural right of access to land. They demanded the distribution of Western lands in order to give inalienable homesteads to poor settlers of all races, including escaped slaves and displaced natives.[101] George Henry Evans, the Free Soil editor of the *Working Man's Advocate*, wrote in an 1844 open letter to the prominent abolitionist Gerrit Smith, "Man has a right on the earth, or he would not be found here. He has a right to exist. He cannot exist without the fruits of the earth. He has a right, therefore, to the fruits of the earth, spontaneous or cultivated."[102] Evans believed that the rights to life, liberty, and the pursuit of happiness are meaningless without the right to soil. In his letter to Smith, he uses a particularly effective extended metaphor to explain how human nature is distorted by the alienation of this right. After noting that 50,000 people received "public charity" during the previous year, he writes:

> Many of these fifty thousand persons have been brought into existence in this city, (which is property of a few gentlemen) just as a bird is brought into existence in a cage .... The bird may chance to make its escape, but it will often return to its cage, not knowing how to seek its natural food .... So it is with the human beings whom the regulations of the landlord master-class have brought into existence on land which they claim as their property.

Evans states that landless "free" workers are their "landlord's slaves," calls Smith "one of the greatest Slaveholders in this country," and demands that he set an example to the world by giving away the vast real estate empire amassed by his father.[103] Astoundingly, Smith did so, and in a letter to the editor of the *Landmark*, he argued that "to abolish chattel-slavery is not to abolish land-monopoly, [but] to abolish land-monopoly is to abolish chattel-slavery."[104] Eventually, Smith gave farms totaling 120,000 acres to almost 2000 free black families, hoping to help them develop the kind of connection to nature that he believed made true freedom possible.[105]

Unlike Smith, some of the politicians associated with the Free Soil Party, especially Van Buren, were both openly racist and content to attempt no more than stop the spread of slavery. Some even made opportunistic appeals to the racism of Northerners voters by arguing that the Wilmot Proviso barred not just slavery, but people of color from the West—free soil was for whites only.[106] Despite the hypocrisy of their electoral representatives, most rank-and-file Free Soilers were active advocates of racial equality who saw the territorial restriction of slavery as a necessary first step to emancipation. Many even openly advocated armed insurrection to overthrow the slavocracy.[107] In other words, as with any broad-based movement, Free Soilers argued urgently over the political decisions that defined the organization and the broader movement. A good deal of energy was wasted comparing the degrees of oppression suffered by wage workers and chattel slaves, and too often Free Soil editors committed outrageous blunders, as when the *Northampton Democrat* asked, "How much worse is it to take a man from his home, than to take a man's home from him?"[108] More often, though, Free Soilers focused on the formal parallels between the situations of wage and slave workers, making clear that they were claiming analytical similarity, not equality of suffering.[109] Horace Greeley, for instance, defined slavery as "that condition in which one human being exists mainly as a convenience for other human beings – in which the time, the exertions, the faculties of part of the Human Family are made to subserve, not their own development, physical, intellectual and moral, but the comfort, advantage or caprices of others."[110]

When they insisted on the fact that the labor of *both* wage workers and slaves was coerced and exploited, Free Soilers were repurposing a common pro-slavery argument, one that was stated by George Fitzhugh in his notorious screeds, *Sociology for the South* (1854) and *Cannibals All!* (1857). But where Fitzhugh and other Southern apologists saw economic hierarchy as a necessary element of "civilization" and advocated the extension of slavery around the world, Free Soilers called for solidarity between wage workers and chattel slaves in the struggle for, as one rank-and-filer put it, "a free soil for humanity, the whole of it, without regard to sex or color."[111] A "fellow-laborer" writing in the *Liberator* under the pseudonym "Truth Teller," argues that "it has pleased God that our liberty is identified with that of the slaves at the South" and calls on workers to "speak out fully and freely."

Were they to follow his lead, Truth Teller imagines that the sound of liberty would flood the land:

> And would that I could hear a voice coming over the tops of the Alleghenies, from the pine forest of Maine, the granite hills of New Hampshire, the green mountains of Vermont, from the fishermen of Marblehead, the cordwainers of Lynn, the ragged hills of Berkshire, "wherever the foot of freeman hath pressed," which, gathering into one burning focus, shall be onward over hill and dale, "Startling the haughty South, / with a deep murmur, / God and our Charter's rights, / Freedom forever."[112]

After all, the vast majority of the hundreds of thousands of black and white, male and female, supporters of the Free Soil Party were ordinary workers—mainly farmers, wage-earners, and artisans—and they believed, as one contemporary commentator wrote, that the "real object of slavery is to depress the wages of the working classes – to withhold from the working community, white or black, the just rights of Labor."[113] Free Soilers saw themselves as proud commoners in a society increasingly dominated by an aristocracy of wealth. Many were committed feminists, adding "Free Hearts and Free Homes" to their list of demands.[114] Above all, as the *Voice of Industry* argued in 1847, Free Soilers were "striving to abolish all Slavery, North, South, East, and West, by securing the means of human Liberty to every son and daughter of humanity."[115]

Douglass met and began to work closely with Gerrit Smith and other Free Soilers soon after he arrived in Rochester. By 1849, he was convinced of the power of Free Soil's platform and asked in the *North Star* for November 9 of that year, "What justice is there in the general government giving away, as it does, the millions upon millions of acres of public lands, to aid soulless railroad corporations to get rich?" Instead the federal government should "multiply the free homes of the people" by giving frontier homesteads to the poor, including free blacks and escaped slaves.[116] Douglass's developing ideas about the connections between nature, freedom, and self-reliance were solidified by his experiences as a busy conductor on the Underground Railroad, sometimes working closely with Harriet Tubman, who made multiple trips to the South, eluding her pursuers through her superior knowledge of the countryside. And during his many meetings with John Brown, Douglass heard about plans to establish abolitionist encampments in the Alleghenies that would eventually serve as mountain strongholds of black power. By 1853, Douglass had become the leading black abolitionist in the United States. His increased prominence was partly a result of the powerful Free Soil ideas that helped him argue, in his "Address of the Colored Convention to the People of the United States," that free and enslaved blacks were "American citizens asserting their rights on their own native soil." Among the many demands enumerated in this document, Douglass included the following: "We ask that the right of pre-emption, enjoyed by all white settlers upon the public lands, shall also be enjoyed by colored settlers; and that the word '*white*' be struck from the pre-emption act."[117] For Douglass in 1853, then, it had become clear that human emancipation required a connection to nature of the most intimate, material, and democratic kind. He summed up his new ideas in the slogan he used as a title when he printed the text of his August 1852 speech before the national Free Soil convention in Pittsburgh: "Let All Soil Be Free Soil."[118]

LANDSCAPES OF REVOLUTION IN *MY BONDAGE* AND *MY FREEDOM*

Douglass's commitment to Free Soil ideas would only grow stronger as the nation moved closer to war. When he selectively revised his *Narrative* in order to produce *My Bondage and My Freedom*, he retained the overall structure of the first book and most of the major incidents. Some sections are only lightly revised, usually in order to amplify political commentary.[119] However, he does intensively rewrite some sections, and they register, at the finest level of figurative detail, his rejection of moralistic for political abolitionism. What he reworks most systematically is his use of landscape *topoi*. In doing so, he moves further along the trajectory established by the journey from the *Narrative* to *The Heroic Slave*. In *My Bondage and My Freedom*, he again transforms Southern nature from an anti-slavery gothic space of paralyzing terror into a revolutionary sublime retreat for awakening to class consciousness. Moreover, he reinvents his own persona, changing himself from an "American slave" into a representative American individual, a self-reliant republican formed by a free childhood in a radical pastoral community of equals.

One of the clearest indications that Douglass was making careful strategic decisions about how to represent nature is an early passage in *My Bondage and My Freedom* in which he lampoons the anti-slavery gothic mode that had been dominant in the *Narrative*. His grandmother sets out on a sorrowful journey to deliver him to the Lloyd plantation and, along the way, they pass through "somber woods":

> She often found me increasing the energy of my grip, and holding her cloth-ing, lest something should come out of the woods and eat me up. Several old logs and stumps imposed upon me, and got themselves taken for wild beasts. I could see their legs, eyes, and ears, or I could see something like eyes, legs, and ears, till I got close enough to them to see that the eyes were knots, washed white with rain, and the legs were broken limbs, and the ears, only ears owing to the point from which they were seen. Thus early I learned that the point from which a thing is viewed is of some importance.[120]

Using wide-eyed monosyllables to parody his breathless fear of the wild and delivering a straight-faced epistemological punchline, Douglass implies that the anti-slavery gothic is a childish fantasy.

Throughout *My Bondage and My Freedom*, Douglass replaces what had been anti-slavery gothic scenes of isolation and privation in the wild with revolutionary sublime scenes of political awakening in nature, perhaps most obviously in his retelling of his struggle with Covey.[121] The autobiographical incidents are fairly straightforward. Covey beat him viciously for falling ill at his work on a hot day, so Douglass ran to his owner, Thomas Auld, to ask for protection. After being rebuffed, he returned to Covey's farm, where he hid in the woods for a full day, trying to decide what to do next. In the *Narrative*, once again, Douglass compresses this day into a single stark sentence: "I spent that day mostly in the woods, having the alternative before me, – to go home and be whipped to death, or stay in the woods and be starved to death."[122] In *My Bondage and My Freedom*, on the other hand, he uses this moment of stasis to establish a clearer motivation for his decision to resist Covey. He begins by sketching the scene: "I am in the wood, buried in its somber gloom, and hushed in its solemn silence; hid from all human eyes; shut in with nature and nature's God, and absent from all human con-trivances. Here was a good place to pray; to pray for help for deliverance – a prayer I had often made before." Here in the divine forest, he is discovered by Sandy Jenkins, "a genuine African," who tells him "with flashing eyes" that "in those very woods, there was an herb" that would prevent Covey from beating him.[123] Of course, Douglass piously dismisses the magical powers of the root, but he nevertheless concludes that Sandy "had been to me the good Samaritan, and had, almost providentially, found me, and helped me when I could not help myself; how did I know but that the hand of the Lord was in it?"[124] Nature has become a school for rebellion in *My*

Bondage and My Freedom, a sacred, anarchic retreat where Douglass and his fellow slaves conspire to resist and even overthrow the power of their masters. Douglass cements his point by concluding the chapter with a quotation from Byron's *Childe Harold's Pilgrimage* that places him squarely in the militant republican tradition: "Hereditary bondmen, know ye not/Who would be free, themselves must strike the blow?"[125]

In addition to deploying the revolutionary sublime, Douglass introduces several elements of the radical pastoral to do the work of self-characterization. In the opening chapter, for example, he focuses close attention on his boyhood home, which is announced in the chapter gloss under the headings "THE LOG CABIN" and "ITS CHARMS."[126] The definite article here emphasizes the conventionality of this trope during the antebellum period, when it served as a shorthand warrant of republican authenticity. Especially after Andrew Jackson's presidential campaigns, a log cabin became as essential to a public life as a clapboard church is now.[127] In the first few pages of *My Bondage and My Freedom,* Douglass describes the charmed domestic setting in which he is raised: his grandparents' family is poor but enjoys a kind of limited autonomy. Describing their house, he plays on the pompous vocabulary of elite estate owners in order to emphasize the hardy poverty of his own origins:

> The dwelling of my grandmother and grandfather had few pretensions. It was a log hut, or cabin, built of clay, wood, and straw. At a distance it resembled – though it was smaller, less commodious and less substantial – the cabins erected in the western states by the first settlers. To my child's eye, however, it was a noble structure, admirably adapted to promote the comforts and conveniences of its inmates.[128]

By professing his humble origins and silently contrasting his cabin with his master's mansion, Douglass validates his credentials as a genuine republican, the natural enemy of all those who would gild their houses (and their language) in order to display their own wealth and power. He winks to acknowledge his participation in political kitsch by referring to the cabin as his "good old home," and then he rhythmically insists on the mutually sustaining connections between modest quarters, pastoral surroundings, and sympathy with humble beings:

> The old cabin, with its rail floor and rail bedsteads upstairs, and its clay floor downstairs, and its dirt chimney, and windowless sides ... was MY HOME – the only home I ever had; and I loved it, and all connected with it. The old

fences around it, and the stumps in the edge of the woods near it, and the squirrels that ran, skipped, and played upon them, were objects of interest and affection. There, too, right at the side of the hut, stood the old well, with its stately and skyward-pointing beam, so aptly placed between the limbs of what had once been a tree, and so nicely balanced that I could move it up and down with only one hand, and could get a drink myself without calling for help.[129]

Harnessing the conventional function of radical pastoralism during the period, Douglass suggests that a wholesome childhood of rural simplicity allowed him to develop the natural sympathies and honest character that now direct his adult commitment to human equality.[130]

RACE AND LABOR IN *MY BONDAGE AND MY FREEDOM*

Now that he has established common ground with his Free Soil readers, Douglass turns to the work of refuting the Southern political theory that the hierarchies of power in slavery reflect natural racial differences. Douglass develops what had been an emergent thread in the *Narrative*: the suggestion that racist ideas are more than just falsehoods that are used to justify racial oppression. They also serve an important secondary political function in Southern society by preventing black and white workers from joining forces to fight for their mutual emancipation from a social order that systematically exploits them both. In response, he develops a class analysis of slavery as a particular sector of American capitalism, and he demonstrates that there are fundamental likenesses between the situations of chattel slaves at the South and wage workers at the North. In doing so, he relies heavily on Free Soil ideas, especially the neo-agrarian idea that class exploitation disturbs an original state of society in which free men maintained their independence by mixing their free labor with the soil. Douglass hoped that these ideas would not only sap the growing ideological power of antebellum racism, but also provide a framework of political theory that could support an alliance between wage and slave labor. *My Bondage and My Freedom*, in other words, is not only an autobiography; it is also a political manifesto that calls for multiracial solidarity to overthrow Southern capitalism and its race-based regime of chattel slavery.[131]

First, with startling candor, Douglass explains that, despite "the supposed hostility of slaveholders to [racial] amalgamation ... *practical* amalgamation is common in every neighborhood where I have been in slavery."[132] More particularly, rape is a daily feature of life in the South, since "by the laws of slavery, children, in all cases, are reduced to the condition of their mothers. This arrangement admits of the greatest license to brutal slave-holders, and their profligate sons, brothers, relations, and friends, and gives to the pleasure of sin, the additional attraction of profit."[133] Because there is an economic incentive that encourages white men to take advantage of their power over black women, the racial binary on which slavery has depended has now dissolved. When Douglass first arrives at Colonel Lloyd's planta-tion, he finds himself "in the midst of a group of [slave] children of many colors; black, brown, copper colored, and nearly white."[134]

When Douglass is sent to St. Michaels to work for Thomas Auld, he finds that he must steal food in order to supplement his meager rations. This occasions a startling meditation on property as theft: "Considering that my labor and person were the property of Master Thomas, and that I was by him deprived of the necessaries of life – necessaries obtained by my own labor – it was easy to deduce the right to supply myself with what was my own."[135] There is a new and critically important emphasis here on accumulated property as appropriated labor value. Auld's food, his farm, all of his resources are either the direct products of his slaves' labor or were amassed indirectly by selling those products and keeping the profits. By refusing to provide adequate food to his slaves, he is attempting to extract an even larger margin of the value of their labor. When they steal food from him, they are simply reappropriating their own stolen labor. Douglass generalizes from his analysis of the economic relations on Auld's farm and argues that it provides a microcosmic model of the South as a whole:

> "I am," thought I, "not only the slave of Thomas, but I am the slave of soci-ety at large. Society at large has bound itself, in form and in fact, to assist Master Thomas in robbing me of my rightful liberty, and of the just reward of my labor; therefore, whatever rights I have against Master Thomas, I have, equally, against those confederated with him in robbing me of liberty."[136]

Suddenly, slavery is no longer a matter of individual conscience, suscepti-ble to change by moral appeals. It is an impersonal social order structured from top to bottom by the definitive relationship of economic exploitation between its two main classes, slaveholders and slaves.

Slavery now begins to sound quite like the system that exists on the other side of the Mason Dixon line. Both are socioeconomic orders dominated by a tiny profiteering elite whose power and wealth depend on the subordination and impoverishment of the vast majority.[137] Whether exploiting wage or slave labor, each system gives "to the lazy and idle, the comforts which God designed should be given solely to the honest laborer."[138] In order to foreground this likeness, Douglass systematically adopts not only the political ideas, but also the vocabulary of Northern working-class radicalism, especially in the chapters describing his years working in the shipyards of Baltimore and delivering his wages to Hugh Auld at the end of every week:

> I was now getting – as I have said – a dollar and fifty cents per day. I contracted for it, worked for it, earned it, collected it; it was paid to me, and it was *rightfully* my own; and yet, upon every returning Saturday night, this money – my own hard earnings, every cent of it – was demanded of me, and taken from me by Master Hugh. He did not earn it; he had no hand in earning it; why, then, should he have it?

Douglass systematically explains his experience as a slave using the lexicon of the wage worker, concluding that his master is a "robber" who compels him "to give him the fruits of my labor, and this power was his only right in the case."[139]

One of the more intensively rewritten sections of *My Bondage and My Freedom* describes how a group of white apprentices brutally beat Douglass in order to establish the color line in the shipyards of Baltimore. After narrating this incident, Douglass adds one of the book's longest passages of direct political commentary, a discussion of the secondary political function of racism in Southern society:

> The slaveholders, with a craftiness peculiar to themselves, by encouraging the enmity of the poor, laboring white man against the blacks, succeeds in making the said white man almost as much a slave as the black slave himself. The difference between the white slave, and the black slave, is this: the latter belongs to *one* slaveholder, and the former belongs to *all* the slaveholders, collectively. The white slave has taken from him, by indirection, what the black slave has taken from him, directly, and without ceremony. Both are plundered, and by the same plunderers. The slave is robbed, by his master, of all his earnings, above what is required for his bare physical necessities; and the white man is robbed by the slave system, of the just results of his labor, because he is flung into competition with a class of laborers who work without wages.[140]

Douglass explains that racism masks "*the conflict of slavery with the interests of the white mechanics and laborers of the south.*"[141] Because slaves are superexploited, given only enough of the produce of their labor to keep them alive for another day of work, their presence in the labor market drives down wages for all: "The slave is robbed, by his master, of all his earnings, above what is required for his bare physical necessities; and the white man is robbed by the slave system, of the just results of his labor, because he is flung into competition with a class of laborers who work without wages."[142] Douglass concludes that simply in order to defend the comparatively generous terms of their own exploitation, Southern wage workers will have to become abolitionists: "competition, and its injurious consequences, will, one day, array the non-slaveholding white people of the slave states, against the slave system, and make them the most effective workers against the great evil." The most significant obstacle to this possibility becoming a reality is racism, which Douglass represents not as a passive belief, but as an active ideology: "At present, the slaveholders blind them to this competition, by keeping alive their prejudice against the slaves, *as men* – not against them *as slaves*." By fomenting racial hatred, the slaveholders "succeed in drawing off the minds of the poor whites from the real fact, that, by the rich slave-master, they are already regarded as but a single remove from equality with the slave."[143] That "single remove" is quite specific. Both slaves and wage workers are deprived of the "just reward" of their labor, but for now only slaves are also robbed of their "rightful liberty."[144] In other words, abolition is a working-class issue, one that both Northern *and* Southern workers should take up not only on principle, but also because the continuing existence of slave labor in the South corrodes working conditions for free labor everywhere.[145]

Since racism is the primary obstacle to solidarity between white workers and black slaves, Douglass takes every opportunity to demonstrate that racial differences are anything but natural. For instance, in a humorous passage near the beginning of *My Bondage and My Freedom*, he makes a characteristic reversal of expectations by claiming that slave children live freer lives than do their young masters:

> The slave-boy escapes many troubles which befall and vex his white brother. He seldom has to listen to lectures on propriety of behavior, or on anything else. He is never chided for handling his little knife and fork improperly or awkwardly, for he uses none. He is never reprimanded for soiling the table-cloth, for he takes his meals on the clay floor. He never has the misfortune, in his games or sports, of soiling or tearing his clothes, for he has almost none to soil or tear. He is never expected to act like a nice little gentleman, for he is only a rude little slave. Thus, freed from all restraint, the slave-boy can be, in his life and conduct, a genuine boy, doing whatever his boyish nature suggests .... His days, when the weather is warm, are spent in the pure, open air, and in the bright sunshine.[146]

A vigorous life of pastoral simplicity, Douglass suggests, allows young slaves, like the common children of the North, to grow freely according to a fundamentally sound human nature. But the Southern elite live lives as remote and artificial as the European aristocracy before them: "Col. Lloyd's plantation resembles what the baronial domains were during the middle ages in Europe. Grim, cold, and unapproachable by all genial influences from communities without, there it stands."[147] The children of slaveholders, like hothouse flowers or overbred livestock, suffer constant disruption and redirection of their innate impulses. Following through on the implications of this comparison, Douglass makes the startling statement that slavery "is a greater evil to the master than to the slave," since "in the case of the slave, the miseries and hardships of his lot are imposed by others, and, in the master's case, they are imposed by himself." Slave-owners commit a kind of moral suicide by perverting their own nature in order to maintain their dominance in a social order based on cruelty and violence. This bold claim finds its clearest expression in a shining apothegm: "The slave is a subject, subjected by others; the slaveholder is a subject, but he is the author of his own subjection."[148]

Douglass's most forceful example of the owners' self-abasement is Mrs. Auld, who changes over time from a kind-hearted to a cruel mistress. This transformation prompts him to observe, "Nature has done almost nothing to prepare men and women to be either slaves or slaveholders. Nothing but rigid training, long persisted in, can perfect the character of the one or the other. One cannot easily forget to love freedom; and it is as hard to cease to respect that natural love in our fellow creatures."[149] This is not moral suasion, designed to convince slave-owners to emancipate their inferior chattels in order to save their own souls. Rather, in response to the pseudo-scientific assertion that nature, specifically inherited racial difference, ordains slavery, Douglass makes the categorically opposite claim: "Nature had made us *friends*; slavery made us *enemies*."[150]

Douglass even goes so far as to parody his own youthful acceptance of racist ideas, making their absurdity explicit: "By some means I learned ... that '*God, up in the sky*,' made every body; and that he made *white* people to be masters and mistresses, and *black* people to be slaves. This did not satisfy me, nor lessen my interest in the subject."[151] Following deadpan irony with brisk seriousness, he exhibits the clearest possible evidence that racist explanations of social hierarchy are simply false: "I found that there were puzzling exceptions to this theory of slavery on both sides, and in the middle. I knew of blacks who were *not* slaves; I knew of whites who were *not* slaveholders; and I knew of persons who were *nearly* white, who were slaves. *Color*, therefore, was a very unsatisfactory basis for slavery." If there is no warrant in nature for oppression, since race is a meaningless category, suddenly the monolith of slavery begins to shudder and sway: "It was not *color*, but *crime*, not *God*, but *man*, that afforded the true explanation of the existence of slavery; nor was I long in finding out another important truth, viz: what man can make, man can unmake." Immediately after narrating his rejection of racial justifications of slavery, Douglass describes the resulting moment of awakening to self-determination:

> I could not have been more than seven or eight years old, when I began to make this subject my study. It was with me in the woods and fields; along the shore of the river, and wherever my boyish wanderings led me; and though I was, at that time, quite ignorant of the existence of the free states, I distinctly remember being, *even then*, most strongly impressed with the idea of being a freeman some day.[152]

This passage closes down Douglass's argument for the moment, and in doing so, it returns to the revolutionary sublime, which is reduced here to a set of generic tags. The symbolic landscape in which he roots his anti-racist argument demonstrates by association that when he determines to make himself free, he is following the promptings of his uncorrupted human nature in the temple of nature.

A Socio-Environmental Theory of Slavery

The first sentences of *My Bondage and My Freedom* announce Douglass's new focus on how slavery, which he calls an "earth-polluting business," simultaneously wreaks social and ecological destruction.[153] Douglass describes Tuckahoe in Talbot Country, Maryland as "remarkable for nothing that I know of more than for the worn-out, sandy, desert-like appearance of its soil, the general dilapidation of its farms and fences, the indigent and spiritless character of its inhabitants, and the prevalence of ague and fever." Just as slavery degrades human beings, so it simultaneously depletes the land. A few sentences later, Douglass repeats the pairing of human and environmental degradation for emphasis: Tuckahoe "is seldom mentioned but with contempt and derision, on account of the barrenness of its soil, and the ignorance, indolence, and poverty of its people." By way of contrast with the slaveholders' careless and destructive agricultural practices,

Douglass relates that his "Grandmother Betty" was "more provident than most of her neighbors in the preservation of seedling sweet potatoes … owing to the exceeding care which she took in preventing the succulent root from getting bruised in the digging, and in placing it beyond the reach of frost, by actually burying it under the hearth of her cabin during the winter months."[154] Her example shows that slaves who are raising crops to supplement their meager rations pay close attention to the needs of the plants because their own physical well-being depends on the health of the soil.

After this introductory contrast between folk agriculture and industrial farming, Douglass systematically deconstructs pro-slavery pastoralism. According to apologists for slavery, the typical plantation was a fertile agricultural landscape that was maintained by the pleasing labor of an extended family who were guided by a benevolent patriarch. Before dismantling that serene image, Douglass gives the reader a moment to contemplate it by describing the first impression that the Lloyd estate made upon him when he was sent there as a young child. The plantation is, "to outward seeming, a most strikingly interesting place, full of life, activity, and spirit." There are "plenty of children to play with" and "a creek to swim in." A windmill presides over a river where sloops carry produce to market. There are "a great many houses," as well as "barns, stables, store-houses, and tobacco-houses; blacksmiths' shops, wheelwrights' shops, [and] coopers' shops."[155] A great bustle of productive economic activity goes on in these buildings and the "kitchens, wash-houses, dairies, summer-house, green-houses, hen-houses, turkey-houses, pigeon-houses, and arbors, of many sizes and devices, all neatly painted, and altogether interspersed with grand old trees, ornamental and primitive, which afforded delightful shade in summer, and imparted to the scene a high degree of stately beauty."[156] Presiding over this lively scene, the Great House stands on a hill, fronted by "a beautiful lawn … dotted thickly over with delightful trees, shrubbery, and flowers" and surrounded by "parks, where – as about the residences of the English nobility – rabbits, deer, and other wild game, might be seen, peering and playing about, with none to molest them or make them afraid." All in all, the plantation is a place of "almost Eden-like beauty [where] the red-winged black-birds [make] all nature vocal with the joyous life and beauty of their wild, warbling notes." The people here live in harmony with one another and with the natural community from which they receive sustenance without labor. Douglass brings this ecstatic paragraph to a sudden close with a crushingly deadpan statement that

reveals how completely spellbound his younger self had been by the scenes
he saw during his first days on the plantation: "These all belonged to me,
as well as to Col. Edward Lloyd, and for a time I greatly enjoyed them."[157]

Next, Douglass tests the plantation pastoral *topos* against reality and
reveals that the Lloyd establishment is a perverse inversion of its bucolic
appearance. It is "a secluded, dark, and out-of-the-way place [that is] seldom visited by a single ray of healthy public sentiment." As a result of this
isolation, "slavery, wrapt in its own congenial, midnight darkness, can, and
does, develop all its malign and shocking characteristics."[158] The children
who play together on the plantation are regarded not as individual beings,
but as a "human crop." The frantic hum of economic activity is driven by
a culture of violence so pervasive that it infects the slaves themselves.
"Everybody, in the south," Douglass writes, "wants the privilege of whipping someone else."[159] Colonel Lloyd whips Old Barney for imagined
faults. When Captain Anthony whips Esther, he is "cruelly deliberate, and
protract(s) the torture, as one who was delighted with the scene."[160] Even
"Uncle Isaac Copper," a disabled slave who leads the children in daily
prayer, beats the inattentive with a stick. This distorted performance of
authority reflects his internalization of the rigid hierarchy of "rank and
station" that extends from Colonel Lloyd at the top, down through subordinate masters like Captain Anthony to overseers like Mr. Sevier, and on
to slaves like Aunt Katy, who are in a position to command the obedience
of others.[161] This cruel social environment distorts the character of everyone who must endure it. "The slaveholder, as well as the slave, is the victim of the slave system [since a] man's character greatly takes its hue and
shape from the form and color of things about him." Captain Anthony, for
instance, "was not by nature worse than other men," but his position
requires him "to be literally insensible to the claims of humanity." As a
result, he has become "an unhappy man" who mutters to himself "as if
defying an army of invisible foes."[162] Douglass is careful to explain that the
slaveholders engage in violence "from motives of policy, rather than from
a hardened nature, or from innate brutality."[163] They are well aware that
"[t]here is no earthly inducement, in the slave's condition, to incite him
to labor faithfully [and that the] fear of punishment is the sole motive for
any sort of industry."[164] Rather than the heads of a loving family, the slaveholders are avaricious tyrants who use violence and torture to force their
slaves to mix their labor with the land in order to extract money.

In order to reinforce the point that slavery is driven by the cold pursuit
of profit, Douglass turns next to refuting "the boast of slaveholders, that

their slaves enjoy more of the physical comforts of life than the peasantry of any country in the world." He states flatly: "My experience contradicts this." Then he details the food, clothing, shelter, and other material conditions of the slaves' daily lives. The "entire monthly allowance of a full grown slave, working constantly in the open field, from morning until night, every day in the month except Sunday [was] a fraction more than a quarter of a pound of meat per day, and less than a peck of corn-meal per week." Often, the meat was "tainted," and much of the corn was "fit only to feed pigs." The clothing provided to the slaves, which consisted of two shirts, two pairs of pants, a pair of shoes, and a jacket, "could not have cost more than eight dollars per year." Most slaves had one cheap blanket and slept on a dirt floor with no privacy. And they were required to "work often as long as they can see" and often had to "take their 'ash cake' with them, and eat it in the field."[165]

After sketching the material privation that the slaves must endure, Douglass turns his eye on the Great House, where "we shall find that height of luxury which is the opposite of that depth of poverty and physical wretchedness that we have just now been contemplating." When slaves are forced "to toil through the fields, in all weathers, with wind and rain beating through [their] tattered garments," their suffering makes it possible for the slaveholders to accumulate a revolting surplus:

> The table groans under the heavy and blood-bought luxuries gathered with painstaking care, at home and abroad. Fields, forests, rivers and seas, are made tributary here. Immense wealth, and its lavish expenditure, fill the great house with all that can please the eye, or tempt the taste. Here, appetite, not food, is the great *desideratum*. Fish, flesh and fowl, are here in profusion. Chickens, of all breeds; ducks, of all kinds, wild and tame, the common, and the huge Muscovite; Guinea fowls, turkeys, geese, and pea fowls, are in their several pens, fat and fatting for the destined vortex. The graceful swan, the mongrels, the black-necked wild goose; partridges, quails, pheasants and pigeons; choice water fowl, with all their strange varieties, are caught in this huge family net. Beef, veal, mutton and venison, of the most select kinds and quality, roll bounteously to this grand consumer. The teeming riches of the Chesapeake bay, its rock, perch, drums, crocus, trout, oysters, crabs, and terrapin, are drawn hither to adorn the glittering table of the great house.[166]

This catalog of luxury foods, a cornucopia of nouns, embodies what it indicates, namely ostentatious consumption well beyond the point of need. Douglass makes clear that this spectacle of gratuitous waste is a show of power. Its central purpose is to demonstrate the "tributary" status both of the slaves who harvest and prepare the food and the land that grows it. Thus, not only does its meaning depend on the simultaneous hunger of slaves, but, adding insult to injury, the labor value expropriated from those slaves is expended in a ritual that is designed to display to them the fact of their own enslavement, to show them that they not only suffer the theft of their labor, but they have also been forcibly alienated from nature. The liberty they have lost is the specific freedom to mix their labor with the land and enjoy the fruits of their industry.[167]

LAND AND BLACK FREEDOM

As the most militant and creative leader of the most successful revolutionary movement of the nineteenth century, Frederick Douglass can serve well as a guide for today's environmental activists. His revolutionary sublime scenes can inspire us to find in the wilderness the determination that we will need as we attempt to radically transform our global society's relationship with the planet. He can help us see that the exploitation of people and of nature are two sides of the same coin in societies that are driven by the pursuit of profit. Just as importantly, he can offer us a vision, rooted in the radical pastoral, of a free human community that possesses the land with care. In a September 1851 speech in Buffalo, New York, Douglass condemned the idea that freed slaves should be resettled in Africa. Instead, he asserted that African Americans had long since established their claim to the land of the free:

> I believe that simultaneously with the landing of the pilgrims, there landed slaves on the shores of this continent, and that for two hundred and thirty years and more we have had a foothold on this continent. We have grown up with you, we have watered your soil with our tears, nourished it with our blood, tilled it with our hard hands. Why should we not stay here? We came when it was a wilderness, and were the pioneers of civilization on this continent. *We* leveled your forests, *our hands* removed the stumps from your fields, and raised the first crops and brought the first produce to your tables. We have been with you, have been with you in adversity, and by the help of God will be with you in prosperity.[168]

The blood and tears that had figured as pollutants in Douglass's letter to Garrison are represented here as the water and nutrients that make soil fertile and productive. For more than two centuries, he argues here, black labor has mixed with nature to nourish and cultivate the growth of a republic that must now acknowledge its debt by granting African Americans freedom and full citizenship. This idea occurs regularly in his writings and addresses during the 1850s. After the Civil War, he developed his thinking about the political significance of black agriculture further when he was invited to speak to the Colored Agricultural and Mechanical Association at their third annual fair in Nashville, Tennessee in September 1873. In this address, Douglass begins by wryly apologizing for speaking on a subject about which he knows so little, since he has worked as a writer, editor, public speaker, and organizer for the past three decades. But he then turns the occasion into an opportunity to talk about how agriculture can be "made to serve us" in "the new order of things." He notes that the freedmen of the South have been excluded from "all respectable employments" by a white population whose sentiment is "let the negro starve!" In response, he articulates a black agrarian program for economic self-sufficiency: "In these circumstances, I hail agriculture as a refuge for the oppressed. The grand old earth has no prejudices against, race, color, or previous condition of servitude."[169] Douglass argues that the hierarchies of power under slavery had reduced the productivity of Southern agriculture. Slaves had no interest in doing their work well. In fact, they had strong motives to impoverish their owners, especially rich masters who relied on cruel drivers to extract the maximum profit from their land and slaves:

> The very soil of your State was cursed with a burning sense of injustice. Slavery was the parent of anger and hate. Your fields could not be lovingly planted nor faithfully cultivated in its presence. The eye of the overseer could not be everywhere, and cornhills could be covered with clods in preference to soft and pulverized soil in their absence, for the hand that planted cared nothing for the harvest. Thus you will see that emancipation has liberated the land as well as the people.

Now that the war was over, though, the South would be "cultivated by liberty" and would enjoy "a vast and general increase of happiness and prosperity."[170] Douglass enjoins his audience to leave behind the culture of violence they had endured and to treat their animals with "uniform sympathy and kindness."[171] He reminds them to take good care of their tools and implements and to "be sure of your water and wood!"[172] He calls on them to "provide for the wants of the soil" and "let nothing be wasted" that might be composted.[173] He encourages them to read "agricultural books and papers" and use natural methods of pest control. Finally, he predicts that by succeeding as farmers, the freedmen of the South will answer whether "the black man will prove a better master to himself than his white master was to him."[174] After all, slavery has left them impoverished and uneducated, but they have the most important thing they need to elevate themselves: "We can work, and the grateful earth yields as readily and bountifully to the touch of black industry as white. We can work, and by this means we can retrieve all our losses."[175] Douglass concludes by urging his listeners to "accumulate property." He acknowledges that this may seem an unscriptural injunction, since we are "accustomed to hear that money is the root of all evil," but he argues that intellectual and spiritual accomplishments must be built on a firm material base, "for without property, there can be no leisure. Without leisure, there can be no thought. Without thought, there can be no invention. Without invention, there can be no progress."[176] Douglass was certainly aware that he was repurposing one of the favorite arguments of apologists for slavery, who claimed that the concentration of wealth and labor of slaves made it possible for the slaveholders to attain a level of cultivation and civilization that was impossible in a republic. But he radicalizes this line of thinking, turning it into a call for self-reliance on the land.

In "Agriculture and Black Progress," which he delivered toward the end of Reconstruction when Southern whites were imposing Jim Crow segregation, Douglass returns to the radical pastoral for renewed inspiration. He encourages his listeners to claim their place in nature and build a free and self-possessed black community that sustains itself by possessing and husbanding the land. Douglass's vision was shared by hundreds of thousands of black farmers who, in the postwar decades, claimed homesteads and built farming communities in the Black Belt, the Midwest, and across the country.[177] The black agrarian movement was carried forward in the early twentieth century by Marcus Garvey's Universal Negro Improvement Association and the National Federation of Colored Farmers. Later, the National Black Farmers Association and the Black Farmers and Agriculturalists Association formed to combat persistent institutional racism in the U.S. Department of Agriculture and in the banks that finance small farmers. In the present, Black Urban Growers and similar organizations are addressing food justice issues by encouraging self-sufficiency. As Kimberly Smith writes, black agrarianism emphasizes "the role of free labor and social and political equality in establishing a sustainable and morally beneficial relationship to the natural world."[178] Rather than seeing "the American landscape ... as pristine and innocent wilderness," this tradition views it as "a corrupted land in need of redemption." Moreover, humans "are to be active, creative, co-equal partners in giving meaning to and redeeming the natural world."[179] Frederick Douglass's radical pastoral vision and the black agrarian tradition that it inspired have much to teach the environmental movement as we adapt to a world in which restoration and reparation are called for more commonly than preservation and conservation.

NOTES

1. The title of this section was provided by the headline of Francie Latour's *Boston Globe* interview with Carolyn Finney. The issue of environmental racism was first brought to public attention by the United Church of Christ Commission for Racial Justice report, *Toxic Wastes and Race*, published in 1987. For a comprehensive survey with several compelling case studies, see Taylor, *Toxic Communities*.
2. Taylor, Grandjean, and Gramann, 9, 11, and 17.
3. Finney, 2–10, provides a comprehensive review of relevant research, especially in the social scientific fields of geography, environmental history, and critical race studies.
4. Quoted in Peterson, 12.
5. Taylor, Grandjean, and Gramann, 17.
6. National Park Service, *A Call to Action*, 9.
7. National Park Service, "Yosemite National Park Ranger."
8. National Park Service, *A Call to Action*, 4.
9. See Myers, 3, for a sharp critique of the National Park Service's history of interpreting canonical "Great Landscape Texts" in ways that enable their segregation by obscuring their active construction and ongoing management, by invoking a common heritage that elides cultural difference, and by foreclosing questions of access. In her memoir about her relationship to her native state of Kentucky, bell hooks writes, "Unmindful of our history of living harmoniously on the land, many contemporary black folks see no value in supporting ecological movements, or see ecology and the struggle to end racism as competing concerns. Recalling the legacy of our ancestors who knew that the way we regard land and nature will determine the level of our self-regard, black people must reclaim a spiritual legacy where we connect our well-being to the well-being of the earth" (39–40).
10. Spence, 4. Mazel set the stage for contemporary studies of "environmentality" by showing how Nature has functioned as "a powerful site for naturalizing constructs of race, class, nationality, and gender" through the mediation of early American literature (xxi). Evans describes the process of

white identity differentiation: "Nature is encountered (and subsequently conquered) by a (white) male figure, who then wrests from the confrontation an instatement or reinstatement of his hegemonic identity. Nature is proffered in these representations as an unproblematic reality, when in fact it is a cultural product designed to serve an ideological function: having conferred upon him his hegemony, Nature is reified as that which has the power to do so" (182). Outka puts this point sharply: "The natural sublime can all too easily serve to 'greenwash' white identity, removing the historical and cultural context that establishes white supremacy, and substituting for it a dehistoricized white individuality and a luminous present moment of fantasized escape from culture, race, and time itself" (24–25). However, as Finseth has shown, Nature also played a key role in anti-slavery and anti-racist discourse in the antebellum era: "Invocations of nature in the cultural fight over the meanings of race entailed a remarkable diversity of representational strategies [and] responded continuously to shifting cultural circumstances" (6). See also Myers, 87–110, for Charles Chesnutt's "resistance to ecological and racial hegemony."

11. Outka, 3 and 80. Michael Bennett observes that the wilderness that is most familiar to black historical experience consists of "trees that were used to enforce southern lynch law." As a result, "anti-pastoralism continues to the present day" as a "main current within African American culture," but it has always "circulated with its dialectical opposite" (207–208). Kimberly Smith summarizes the ways in which the agricultural, legal, political, and folk cultural practices of the South "created a complex relationship between slaves and the landscape, forcing most of its victims into an intimacy with the immediate natural world but also, in some respects, alienating them from it" (14). Above all, "slavery deprived slaves of property rights and the right to travel" and "these rights would become critical components of the freedom sought by black Americans, linking self-possession with possession of the land" (37).

12. Glave, 8. Finney, 10. Robert Butler argues that black writers have "usually found it inappropriate to envision idealized non-urban space as a relief from the pressures of urban living" because of "the historical experience of black people in America" (11).

13. Dixon, 3.

14. The segregation of the environmental commons can be seen as part of the larger pattern of segregation of public space in general. And at a time when countless people are being killed for being black in public, integration of the commons is a fundamental civil rights issue.

15. Speth and Thompson, n.p.

16. Klein, n.p.

17. Alston, "The Summit."

18. Greening Youth Foundation, "The Company."
19. National Park Service, "Grand Canyon Hosts Camping 101."
20. Green 2.0, "Rue Mapp."
21. Diverse Environmental Leaders Speakers Bureau. Two other notable urban environmental justice organizations are Sustainable South Bronx and Alternatives for Community and Environment in Boston.
22. As Bullard documents, activists of color have been engaged in vital environmental justice campaigns for decades. For additional anecdotes and case studies, see Purdy and the first section of Adamson, Evans, and Stein.
23. Buell, *Future*, 119. Reed frames a general research program for "environmental justice ecocriticism." Myers envisions an ambitious interdisciplinary effort to "make ecology a site upon which an egalitarian racial paradigm can be grounded" by integrating "critical race studies and ecocriticism" (8).
24. Glave and Stoll collect a set of essays that represent the impressive range and depth of the burgeoning field of African American environmental history.
25. Deming and Savoy's 2002 collection has a broader focus, both in terms of genre and in terms of cultural heritage: "If what is called 'nature writing' aims to understand how we comprehend and then live responsibly in the world, then it must recognize the legacies of the Americas' past in ways that are mindful of the complex historical and cultural dynamics that have shaped us all. Perhaps some would say this isn't the goal of writing about nature or natural history. But if such writing examines human perceptions and experiences of nature, if an intimacy with and response to the larger-than-human world define who or what we are, if we as people are part of nature, then the experiences of all people on this land are necessary stories, even if some voices have been silent, silenced, or simply not recognized as nature writing. What is defined by some as an edge of separation between nature and culture, people and place, is a zone of exchange where finding common ground is more than possible; it is necessary" (6–7).
26. Dungy, xxi.
27. Ruffin, 2–3 and 16.
28. Ruffin, 25–55. Focusing on the work of poet and critic Sterling Brown, Anderson argues that African American literature has relied heavily on the georgic mode, which emphasizes the moral value of agricultural labor, both to reframe black relations with nature and to critique the slavocrats' plantation pastoral fantasies. Ronda makes a related argument, showing that Paul Lawrence Dunbar's georgic poems in *Lyrics of Lowly Life* "disarticulate manual labor from discourses of racial self-improvement and emancipation" at the same time that they "assert the sympathetic humanity and blamelessness of African Americans in the face of virulent institutional racism" (864). See also Montrie's chapter, "Living by Themselves: Slaves'

and Freedmen's Hunting, Fishing, and Gardening in the Mississippi Delta" (35–52), in *Making a Living*.

29. Ruffin, 6. Kimberly Smith carries out important historical reconstruction in *African American Environmental Thought* (2007), in which she describes a long tradition of theory and practice that has not been recognized as valuable by mainstream environmentalism because "African American activists [have] usually framed their concerns as civil rights issues" (3). Smith argues that "they *are* civil rights issues based on the assumption that environmental amenities and freedom from environmental harms are critical to the good life and should be available to all" (7–8, emphasis added). She shows that "black theorists reasoned that race slavery and post-Emancipation racial oppression put black Americans into a conflicted relationship with the land—by coercing their labor, restricting their ability to own land, and impairing their ability to interpret the landscape." As a result, a "central theme in this tradition is the claim that denial of freedom to black Americans has distorted their relationship to the natural environment; indeed, it has scarred the land itself. To black writers working in this tradition, America—not just the political community but the physical terrain—is a land cursed by injustice and in need of redemption." On the other hand, "plantation slaves experienced the American environment in the context of a struggle to achieve self-mastery through mastery of a disordered and often hostile natural world. That struggle made the connection between nature and freedom—between possessing the land and possessing oneself—a central theme in black political thought" (19).

30. Black abolitionism was an international movement whose transatlantic networks have been incompletely mapped. Gilroy traces the contours of a black Atlantic culture that transcends national boundaries and literary-historical periods. Delbanco tells the riveting story of the fugitive slaves, who, during the antebellum decades in the United States, "pushed the nation toward confronting the truth about itself" and "incited conflict in the streets, the courts, the press, the halls of Congress, and perhaps most important in the minds and hearts of Americans who had been oblivious to their plight" (2). Michael Bennett's *Democratic Discourses* demonstrates how radical abolitionists, particularly black abolitionists, transformed political consciousness and literary culture by insisting that "one cannot think about democracy in the United States without thinking about gender, race, and class" (8).

31. Ball, 326.

32. Ball, 328.

33. Ball, 333.

34. Ball, 336. The second half of Ball's *Slavery in the United States* also relies heavily on the radical pastoral *topos*. Ball escapes from Georgia and makes

his way across country to Pennsylvania. He lives off the land for several months, surviving intense privation, but he is still able to coolly observe daily life on the plantations that he passes.

35. Marrant, 9 and 14.
36. Marrant, 15.
37. Marrant, iv.
38. Marrant, 21.
39. Marrant, 28.
40. Marrant, 30 and 29.
41. Marrant, 31–32.
42. Marrant, 39.
43. Equiano, 25.
44. Equiano, 8.
45. Equiano, 3 and 13–14.
46. Equiano, 8 and 14.
47. Equiano, 12 and 14.
48. Equiano, 4–5 and 7.
49. Equiano, 57 and 143.
50. Parts of the following section were published as Newman, "Free Soil and the Abolitionist Forests of Frederick Douglass's 'The Heroic Slave.'"
51. Jay argues that Douglass "had no choice but to draw upon the cultural archive of white society and then transform his materials rhetorically." In doing so, he achieved a "mastery of the master's tongue [that transformed] him from the dictated subject of ideology into the agent of historical (and literary) change" (224 and 228).
52. My reading of the landscapes of *My Bondage and My Freedom* differs from Joseph Bodziock's in "The Cage of Obscene Birds: The Myth of the Southern Garden in Frederick Douglass's *My Bondage and My Freedom*." Bodziock sees the *Narrative* as minimalist, and argues that the second autobiography "used Gothic modes of expression" to invert the "moon-and-magnolias" myth peddled by apologists for slavery and to portray the South instead as a corrupt garden occupied by a decadent aristocracy. This misalignment may reflect the fact that Bodziock focuses mainly on the characterization of figures like Covey and Lloyd, while my own reading is more concerned with passages that describe the land.
53. In "Violence Done to Nature," Finley notes that most historians of Free Soil emphasize the racism of the movement's political candidates and ignore the broadly multiracial character of its rank-and-file participants (5). Finley analyzes "the archive of antislavery literature that treats slavery as unnatural and as an ecological problem" and shows that when "Free Soil doctrine is refashioned into the literature of Free Soil, its imagery, tropes, heuristics, and presumptions become more imaginative, evocative,

and affecting, as well as increasingly sophisticated and complex, enabling not only a more incisive and radical condemnation of slaveholding hegemony, slaveholding practices, slavery's effects on land, and anti-black racism across the United States, but also a sophisticated critique of the racism of normative Free Soil ideology, its insufficient stance on immediate abolition, and the anti-ecological paradoxes contained within its argument that slavery is pollutive and degrading, yet can be contained to its present location" (4 and 7).

54. Frederick Douglass, *Papers, Series Two* 1.80. Subsequent citations will identify this two-volume series as *PS2*.

55. Douglass, *PS2*, v. 2, 10.

56. Douglass, *PS2*, v. 2, 11.

57. Douglass, *PS2*, v. 2, 11. Later, Douglass accused himself of a "slavish adoration of [his] Boston friends" (*PS2*, v. 2, 227). On the other hand, Garrison never proclaimed an absolute prohibition against politics, even when the split between nonresistant abolitionists and political anti-slavery advocates grew bitter. Moreover, despite his claim to the contrary, Douglass was never a doctrinaire Garrisonian (Sewell, *Ballots*, 24–41.) During his years with the Massachusetts Anti-Slavery Society, he advocated moral suasion not on abstract principle but because he was convinced that it was the strategy that was most likely to produce swift emancipation (Goldstein, 61–72). Also, he was a disciplined organizer who understood the importance of consistency in a group's message.

58. Douglass, *PS2*, v. 1, 20–21.

59. Douglass, *PS2*, v. 1, 40.

60. Douglass, *PS2*, v. 1, 61–62.

61. Douglass, *PS2*, v. 1, 47.

62. Douglass, *PS2*, v. 1, 46.

63. Douglass, *PS2*, v. 1, 48.

64. Douglass, *PS2*, v. 1, 52–53.

65. Riss argues that, in the discourse of liberalism, the meaning of "personhood" has varied radically through history and therefore "questions about the reality of the 'person'" should be "replaced by questions about who controls the terms that establish this conceptual category" (23). Focusing on the fight with Covey, Riss argues that while the *Narrative* is often read as the paradigmatic liberal text, since it assumes the "always already existing 'personhood' of the slave," it simultaneously dramatizes how personhood is "an identity that must be enacted" (167).

66. Howard Jones narrates the events on the *Creole* that inspired *The Heroic Slave*. See Harrold for an account of abolitionist responses to the mutiny. Even Garrison compares Douglass to Patrick Henry in his preface to the *Narrative* (Douglass, *Papers, Series One*, v. 1, 4). Subsequent citations will identify this two-volume series as *PS1*.

67. Andrews argues that Douglass engages in an act of generic radicalization by abandoning the highly managed slave narrative form for the more potentially disruptive historical romance, where he represents Washington's exploits in "fictive, not merely fictional" terms, thus calling specific attention to the way that his text simultaneously dramatizes the factual material on which it is based and makes a claim of representative accuracy (23–30).

68. There has been limited scholarly discussion of *The Heroic Slave*, most of which focuses on its relation to the ideology of nationalism. Walter points out that Douglass's use of republican discourse risks reproducing its racism and sexism. Hopper argues that republicanism functions as part of a rhetoric of universality that attempts to transcend race. Sale concludes that Douglass transforms the discourse in the act of repurposing it (700–704). Finally, Wilson argues that Douglass emphasizes the transnationalism of his hero in order to inoculate the book against jingoism.

69. Douglass, *The Heroic Slave*, 175.

70. Douglass, *The Heroic Slave*, 176–177.

71. Douglass, *The Heroic Slave*, 178, emphasis in original.

72. Douglass, *The Heroic Slave*, 180.

73. Douglass, *The Heroic Slave*, 181–182.

74. Douglass, *The Heroic Slave*, 181

75. Douglass, *The Heroic Slave*, 190.

76. Douglass, *The Heroic Slave*, 190 and 192.

77. Douglass, *The Heroic Slave*, 192–193. Yarborough argues that Madison Washington is characterized according to antebellum notions of bourgeois masculinity, compounding Emersonian self-reliance with such characteristics as "nobility, intelligence, strength, articulateness, loyalty, virtue, rationality, courage, self-control, courtliness, honesty, and physical attractiveness as defined [ironically] in white Western European terms" (168).

78. Douglass, *The Heroic Slave*, 193.

79. Douglass, *The Heroic Slave*, 194.

80. Another example of the revolutionary sublime in early African American fiction is Martin Delany's *Blake; or, the Huts of America*, an 1859 novel about a fugitive slave who takes refuge in the backcountry with a band of Choctaw Indians and is convinced by their chief to return to his plantation and lead an insurrection. Henry finds the determination to pursue liberty through armed rebellion while "standing upon a high bank of a stream, contemplating his mission" in the wilderness (69).

81. Douglass, *The Heroic Slave*, 196.

82. Douglass, *The Heroic Slave*, 198.

83. Douglass, *The Heroic Slave*, 205.

84. For an excellent introduction to the historiography of this topic, see Foner and Shapiro, ix-xxx. This volume gathers a representative sample of pri-

mary sources, mainly editorial columns and correspondence from radical newspapers, that illuminate the interrelations between the abolitionist, utopian socialist, and labor movements of the 1840s and 1850s. For a survey of the diversity of the Northern working class, see Jacqueline Jones, *Social History*, 89–117.

85. Douglass, *PS1*, v. 1, 433–434.
86. Bradbury, 169–186. McFeely, 138–143.
87. Giles, 781.
88. Douglass, *PS2*, v. 2, 217.
89. Douglass, *PS2*, v. 2, 212.
90. Douglass, *PS2*, v. 2, 212.
91. Ferreira, Fulkerson, and Rice and Crawford provide varying accounts of Douglass's transformative exile and return.
92. Foner, 2, 48–66.
93. Bradbury, 171.
94. Douglass, *PS1*, v. 1, 370–371.
95. Colaiaco provides a full-length study of Douglass's republicanism. For a concise survey of republican ideas in the discourse of abolition, see McInerney, 7–25.
96. Quarles, 183–185.
97. Kraut, 71–99, and Sewell, *Ballots*, 43–106.
98. Sewell argues that "antislavery politicians, recognizing the racist nature of their society, conceded the need to float abolition in a larger vessel: the freedom of *all* men from monopoly and class legislation" (*Ballots*, 115).
99. Quarles, 185–187.
100. Earle provides an excellent full-length analysis of the Free Soil movement that emphasizes its ideological diversity. For a broader history of working-class republicanism, see Wilentz, especially 299–396.
101. Lause narrates the history of the antebellum land reform movement and its intimate connections with abolition, socialism, Free Soil, and other radical movements. Stauffer shows that Free Soilers "identified with the Indian as a symbol of the savage fighter *par excellence* who rejected white laws and civilization and [who] found hope, strength, and courage in the wilderness and the Great Spirit in Nature" (237).
102. Quoted in Commons, 353.
103. Quoted in Commons, 354–355.
104. Quoted in Earle, 36.
105. Dyson describes Smith's land distribution efforts. In *This Radical Land*, Daegan Miller narrates this episode of African American agricultural settlement in the Adirondacks as an example of "the ecology of freedom" (47–96).
106. For an account of the growth of black working-class communities and the racist backlash against them in the antebellum North, see Jacqueline

Jones, *American Work*, 141–68 and 246–72. Roediger surveys working-class white-supremacist ideology during the antebellum period, with a special focus on language and popular culture (43–166). Roediger's afterword to the 1999 revised edition includes several important clarifications, partly in response to Lott, who offers a more consistently dialectical analysis of what he calls the "conflicted intimacy of American racial cultures" (75). Consciously anti-racist abolitionism was a powerful, perhaps even dominant, strain in the Free Soil milieu, as documented in Lause, 72–84. Mandel, 61–110, remains a balanced narrative of Northern working-class abolitionism.

107. Sewell, "Slavery."
108. Quoted in Foner and Shapiro, 17.
109. Cunliffe, 1–31, provides a fascinating survey of the transatlantic discourse of "white slavery."
110. Greeley, 353.
111. Quoted in Lause, 80.
112. Quoted in Foner and Shapiro, 72 and 76.
113. West, n.p.
114. Pierson, 71–96.
115. Quoted in Foner and Shapiro, 11. On black leadership of the abolition movement, see Quarles 23–41. Jeffrey, 161–170, surveys women's participation in the Liberty and Free Soil parties. Magdol analyzes the demographics of abolitionism just prior to the split into moral suasion and political wings.
116. Quoted in Foner, 2.13.
117. Douglass, *PS1*, v. 2, 255 and 257.
118. Douglass, *PS1*, v. 2, 388.
119. Douglass reproduces substantial passages of the *Narrative* inside quotation marks, observing that he sees no need for improvement.
120. Douglass, *PS2*, v. 2, 28.
121. My reading of *My Bondage and My Freedom* shares many features with Finseth, who argues that the book's "pastoral ethic … calls into question not only the practices of the South, but also the values of the whole country" and that Douglass "reconceptualized the role of the natural environment in his and his cultures' (plural) experience, coming to see the natural world, both symbolically and literally, as the foundation for a revitalized African American community" (280 and 272).
122. Douglass, *PS2*, v. 1, 52.
123. Douglass, *PS2*, v. 2, 134–135.
124. Douglass, *PS2*, v. 2, 136–137.
125. Douglass, *PS2*, v. 2, 142.
126. Douglass, *PS2*, v. 2, 21, emphasis in original.

127. The log cabin as badge of anti-elitism became so widely familiar in the antebellum years that it could be deployed with no regard whatsoever for autobiographical fact and actual class status. For instance, Nathaniel Hawthorne's 1852 campaign autobiography of his Bowdoin chum, the wealthy lawyer Franklin Pierce, treats the reader to a bait-and-switch ruse in order to introduce the indispensable cabin. Half of Hawthorne's first chapter sketches a portrait of his subject's father, Benjamin Pierce. Hawthorne insists that the elder Pierce grew up with the "simple fare, hard labor, and scanty education as usually fell to the lot of a New England yeoman's family" and that he traded his plow for a gun on the first day of the Revolution. After the victory of the former colonies, he betook "himself to the wilderness for a subsistence" and "built himself a log hut" (8–9). Hawthorne now quietly makes the switch, remarking that from "infancy upward [Franklin Pierce] had before his eyes, as the model on which he might instinctively form himself" his father, whose frontier upbringing ensured that he was "a most decided democrat, and supporter of Jefferson and Madison; a practical farmer, moreover, not rich, but independent ... a man of the people, but whose natural qualities inevitably made him a leader among them" (10). In order to establish his candidate's republican authenticity, Hawthorne's uses the trope of the log cabin to make an invidious comparison between economically and morally self-reliant republicans and a decadent and debased elite. It was irrelevant that Pierce was in fact the Bowdoin-educated scion of one of the oldest families in Massachusetts.
128. Douglass, *PS2*, v. 2, 23.
129. Douglass, *PS2*, v. 2, 25 and 27.
130. Outka reads Douglass's description of his log cabin childhood as "a sort of Blakean Song of Innocence, a stereotypical pastoral nature that depends on ignorance of the realities of the commodified historical and economic landscape in which the child has already been emplaced" (63). When Douglass realizes that he is a slave, his innocence collapses traumatically. Rather than read the passage as a transparent record of traumatic experience, I emphasize the rhetorical virtuosity of Douglass's self-fashioning.
131. Ellis examines the landscapes of *My Bondage and My Freedom* and argues that Douglass raises the "pragmatic objection that slavery may be economically and environmentally hazardous" and "develops a pointedly amoral critique of slavery on the grounds of its practical unsustainability" (276–277).
132. Douglass, *PS2*, v. 2, 67.
133. Douglass, *PS2*, v. 2, 35.
134. Douglass, *PS2*, v. 2, 29.
135. Douglass, *PS2*, v. 2, 108.

136. Douglass, *PS2*, v. 2, 109.

137. Shore, 16–41, describes how the Southern elite countered Smithian critiques of slavery by adopting the lexicon of political economy and describing themselves as "slaveholding capitalists."

138. Douglass, *PS2*, v. 2, 144.

139. Douglass, *PS2*, v. 2, 182.

140. Douglass, *PS2*, v. 2, 177.

141. Douglass, *PS2*, v. 2, 176–177, emphasis in original.

142. Douglass, *PS2*, v. 2, xx.

143. Douglass, *PS2*, v. 2, 177.

144. Douglass, *PS2*, v. 2, 109.

145. Shore, 42–78, gives an important account of the antebellum period's increasingly open class conflict between slaveholding capitalists and the Southern working class.

146. Douglass, *PS2*, v. 2, 25.

147. Douglass, *PS2*, v. 2, 38.

148. Douglass, *PS2*, v. 2, 62.

149. Douglass, *PS2*, v. 2, 87.

150. Douglass, *PS2*, v. 2, 92.

151. Douglass, *PS2*, v. 2, 52.

152. Douglass, *PS2*, v. 2, 53.

153. In this section, I rely heavily on the work of James Finley, who writes: "In his second autobiography, *My Bondage and My Freedom* (1855), Frederick Douglass portrays slavery as an 'earth-polluting business,' an environmentally destructive and ecologically unsustainable system of violence against people and landscapes. Drawing upon natural science, evidence of environmental degradation, and agrarian rhetoric throughout *My Bondage and My Freedom*, Douglass constructs an explicitly ecological antislavery critique. Whereas Douglass's 1845 *Narrative* positions chattel slavery as an unspeakable sin and pressing moral issue, *My Bondage and My Freedom* also renders slavery as deeply unnatural, condemning it as 'violence done to nature' (114), and providing an incisive analysis of a white-supremacist system that animalizes or bestializes people of color and violently manipulates the species-line to maintain racial hegemony while simultaneously blighting and destroying landscapes and natural resources" (8).

154. Douglass, *PS2*, v. 2, 22–23.

155. Douglass, *PS2*, v. 2, 38–39.

156. Douglass, *PS2*, v. 2, 39–40.

157. Douglass, *PS2*, v. 2, 40.

158. Douglass, *PS2*, v. 2, 37.

159. Douglass, *PS2*, v. 2, 41 and 42.

160. Douglass, *PS2*, v. 2, 51.
161. Douglass, *PS2*, v. 2, 46.
162. Douglass, *PS2*, v. 2, 47.
163. Douglass, *PS2*, v. 2, 50.
164. Douglass, *PS2*, v. 2, 123.
165. Douglass, *PS2*, v. 2, 59–60. An ash cake is a lump of corn-meal dough baked directly in the coals of a fire.
166. Douglass, *PS2*, v. 2, 62–63.
167. In *Blood and Earth: Modern Slavery, Ecocide, and the Secret to Saving the World*, Kevin Bales demonstrates that Douglass's critique still holds true. Across the world, criminals and criminal corporations rely on slave labor to reduce costs, with the result that some 38.5 million "slaves are producing many of the things we buy, and in the process they are forced to destroy our shared environment, increase global warming, and wipe out protected species" (8).
168. Douglass, *PS2*, v. 2, 340.
169. Douglass, *PS1*, v. 4, 385–386.
170. Douglass, *PS1*, v. 4, 387.
171. Douglass, *PS1*, v. 4, 388.
172. Douglass, *PS1*, v. 4, 389.
173. Douglass, *PS1*, v. 4, 390.
174. Douglass, *PS1*, v. 4, 393.
175. Douglass, *PS1*, v. 4, 394.
176. Douglass, *PS1*, v. 4, 393.
177. Kimberly Smith, 39–97, surveys black agrarianism in the nineteenth century and offers insightful commentary on the tradition's emphasis on private property ownership as the basis for economic independence and political freedom. As Smith observes, this emphasis runs counter to the mainstream environmental movement's emphasis on public land management.
178. Kimberly Smith, 40.
179. Kimberly Smith, 8.

BIBLIOGRAPHY

Adamson, Joni, Mei Mei Evans, and Rachel Stein, eds. 2002. *The Environmental Justice Reader: Politics, Poetics, and Pedagogy*. Tucson: University of Arizona Press.

Alston, Dana. n.d. The Summit: Transforming a Movement. *Reimagine*. Race, Poverty, and the Environment. http://www.reimaginerpe.org/20years/alston

Anderson, David R. 2016. Sterling Brown and the Georgic Tradition in African-American Literature. *Green Letters: Studies in Ecocriticism* 20 (1): 86–96.

Andrews, William. 1990. The Novelization of Voice in Early African American Narrative. *PMLA* 105 (1): 23–36.

Armbruster, Karla, and Kathleen R. Wallace, eds. 2001. *Beyond Nature Writing: Expanding the Boundaries of Ecocriticism.* Charlottesville: University Press of Virginia.

Bales, Kevin. 2016. *Blood and Earth: Modern Slavery, Ecocide, and the Secret to Saving the World.* New York: Spiegel and Grau.

Ball, Charles. 1837. *Slavery in the United States: A Narrative of the Life and Adventures of Charles Ball, a Black Man.* New York: John S. Taylor. Internet Archive.

Bennett, Michael. 2005. *Democratic Discourses: The Radical Abolition Movement and Antebellum American Literature.* New Brunswick: Rutgers University Press.

———. Anti-pastoralism, Frederick Douglass, and the Nature of Slavery. In ed. Armbruster and Wallace, 195–210.

Bodziock, Joseph. 2004. The Cage of Obscene Birds: The Myth of the Southern Garden in Frederick Douglass's *My Bondage and My Freedom*. In *The Gothic Other: Racial and Social Constructions in the Literary Imagination,* ed. Ruth Bienstock Anolik and Douglas L. Howard. Jefferson: McFarland.

Bradbury, Richard. 1999. Frederick Douglass and the Chartists. In *Liberating Sojourn: Frederick Douglass and Transatlantic Reform,* ed. Alan J. Rice and Martin Crawford, 169–186. Athens: University of Georgia Press.

Buell, Lawrence. 2005. *The Future of Environmental Criticism: Environmental Crisis and Literary Imagination.* London: Blackwell.

Bullard, Robert D. 1993. *Confronting Environmental Racism: Voices from the Grassroots.* Boston: South End.

Butler, Robert. 1995. The City as Liberating Space in *Life and Times of Frederick Douglass.* In *The City in African-American Literature,* ed. Yoshinobu Hakutani and Robert Butler, 21–36. Madison: Fairleigh Dickinson University Press.

Colaiaco, James A. 2006. *Frederick Douglass and the Fourth of July.* New York: Palgrave Macmillan.

Commons, John R., et al. 1910. *A Documentary History of American Industrial Society: Labor Movement, 1840–1860.* Vol. 7. Cleveland: Clark. *GoogleBooks.*

Cunliffe, Marcus. 1979. *Chattel Slavery and Wage Slavery: The Anglo-American Context, 1830–1860.* Athens: University of Georgia Press.

Delany, Martin. *Blake; or, the Huts of America.* In *Uncle Tom's Cabin and American Culture: A Multi-Media Archive,* ed. Steven Railton. http://utc.iath. virginia.edu

Delbanco, Andrew. 2018. *The War Before the War: Fugitive Slaves and the Struggle for America's Soul from the Revolution to the Civil War.* New York: Penguin.

Deming, Alison Hawthorne, and Lauret E. Savoy, eds. 2002. *Colors of Nature: Culture, Identity, and the Natural World.* Minneapolis: Milkweed.

Diverse Environmental Leaders Speakers Bureau. n.d. http://delnsb.com/

Dixon, Melvin. 1987. *Ride Out the Wilderness: Geography and Identity in Afro-American Literature.* Urbana: University of Illinois Press.

Douglass, Frederick. 1853. The Heroic Slave. In *Autographs for Freedom*, ed. Julia Griffiths. Boston: Jewett. Internet Archive.
———. 1979–1992. *Papers. Series One: Speeches, Debates, Interviews*, ed. John W. Blassingame, 5 vols. New Haven: Yale University Press.
———. 1999–. *Papers. Series Two: Autobiographical Writings*, ed. John W. Blassingame, John R. McKivigan, and Peter P. Hinks, 2 vols. to date. New Haven: Yale University Press.
Dungy, Camille, ed. 2009. *Black Nature: Four Centuries of African American Nature Poetry*. Athens: University of Georgia Press.
Dyson, Zita. 1918. Gerrit Smith's Efforts in Behalf of the Negroes in New York. *Journal of Negro History* 3: 354–359.
Earle, Jonathan. 2004. *Jacksonian Antislavery and the Politics of Free Soil, 1824–1854*. Chapel Hill: University of North Carolina Press.
Ellis, Cristin. 2014. Amoral Abolitionism: Frederick Douglass and the Environmental Case Against Slavery. *American Literature* 86 (2): 275–303.
Equiano, Olaudah. 1790. *The Interesting Narrative of the Life of Olaudah Equiano, or Gustavus Vassa, the African*, 3rd ed. London. Internet Archive.
Evans, Mei Mei. 'Nature' and Environmental Justice. In ed. Adamson, Evans, and Stein, 181–193.
Ferreira, Patricia J. 2001. Frederick Douglass and the 1846 Dublin Edition of His Narrative. *New Hibernia Review* 5: 53–67.
Finley, James. 2014. "Violence Done to Nature": Free Soil and the Environment in Antebellum Antislavery Literature. Unpublished Doctoral Dissertation, University of New Hampshire.
Finney, Carolyn. 2014. *Black Faces, White Spaces: Reimagining the Relationship of African Americans to the Great Outdoors*. Chapel Hill: University of North Carolina Press.
Finseth, Ian Frederick. 2009. *Shades of Green: Visions of Nature in the Literature of American Slavery, 1770–1860*. Athens: University of Georgia Press.
Fitzhugh, George. 1857. *Cannibals All! or, Slaves Without Masters*. Richmond: A. Morris. Internet Archive.
———. 1854. *Sociology for the South*. Richmond: A. Morris. Internet Archive.
Foner, Philip S. 1950. *The Life and Writings of Frederick Douglass*, 5 vols. New York: International Publishers.
Foner, Philip, and Herbert Shapiro, eds. 1994. *Northern Labor and Antislavery: A Documentary History*. Westport: Greenwood.
Fulkerson, Gerald. 1974. Exile as Emergence: Frederick Douglass in Great Britain, 1845–1847. *Quarterly Journal of Speech* 60: 69–82.
Giles, Paul. 2001. Narrative Reversals and Power Exchanges: Frederick Douglass and British Culture. *American Literature* 73: 779–810.
Gilroy, Paul. 1993. *The Black Atlantic: Modernity and Double Consciousness*. Cambridge, MA: Harvard University Press.

Glave, Dianne D. 2010. *Rooted in the Earth: Reclaiming the African American Environmental Heritage*. Chicago: Lawrence Hill.

Glave, Dianne D., and Mark Stoll, eds. 2006. *To Love the Wind and the Rain: African Americans and Environmental History*. Pittsburgh: University of Pittsburgh Press.

Goldstein, Leslie Friedman. 1976. Violence as an Instrument for Social Change: The Views of Frederick Douglass (1817–1895). *The Journal of Negro History* 61: 61–72.

Greeley, Horace. 1850. *Hints Toward Reforms, in Lectures, Addresses, and Other Writings*. New York: Harper and Brothers. *GoogleBooks*.

Green 2.0. Rue Mapp. https://www.diversegreen.org/people/rue-mapp/

Greening Youth Foundation. n.d. The Company. http://www.gyfoundation.org/the-company/

Harrold, Stanley. 1999. Romanticizing Slave Revolt: Madison Washington, the *Creole* Mutiny, and Abolitionist Celebration of Violent Means. In *Antislavery Violence: Sectional, Racial, and Cultural Conflict in Antebellum America*, ed. John R. McKivigan and Stanley Harrold, 89–107. Knoxville: University of Tennessee Press.

Hawthorne, Nathaniel. 1852. *The Life of Franklin Pierce*. Boston: Ticknor and Fields. *Google Books*.

hooks, bell. 2009. *Belonging: A Culture of Place*. New York: Routledge.

Hopper, Briallen. 2001. The Bondage of Race and the Freedom of Transcendence in Frederick Douglass's *My Bondage and My Freedom*. *Postgraduate English* 4, September.

Jay, Gregory. 1990. American Literature and the New Historicism: The Example of Frederick Douglass. *Boundary* 2: 211–242. *JSTOR*.

Jeffrey, Julie R. 1998. *The Great Silent Army of Abolitionism: Ordinary Women in the Antislavery Movement*. Chapel Hill: University of North Carolina Press.

Jones, Howard. 1975. The Peculiar Institution and National Honor: The Case of the *Creole* Slave Revolt. *Civil War History* 21: 28–50.

Jones, Jacqueline. 1998. *American Work: Four Centuries of Black and White Labor*. New York: Norton.

———. 1999. *Social History of the Laboring Classes: From Colonial Times to the Present*. Malden: Blackwell.

Klein, Naomi. 2014. Why #BlackLivesMatter Should Transform Climate Debate. *The Nation*, December 12.

Kraut, Alan M., ed. 1983. *Crusaders and Compromisers: Essays on the Relationship of the Antislavery Struggle to the Antebellum Party System*. Westport: Greenwood.

Latour, Francie. 2014. Hiking While Black: The Untold Story. *Boston Globe*, June 20. https://www.bostonglobe.com/ideas/2014/06/20/hiking-while-black-the-untold-story-black-people-great-outdoors/ssRvXFYogkZs2e4RX3z6JP/story.html

Lause, Mark A. 2005. *Young America: Land, Labor, and the Republican Community.* Urbana: University of Illinois Press.

Lott, Eric. 1993. *Love and Theft: Blackface Minstrelsy and the American Working Class.* New York: Oxford University Press.

Magdol, Edward. A Window on the Abolitionist Constituency: Antislavery Petitions, 1836–1839. In ed. Kraut, 45–70.

Mandel, Bernard. 1955. *Labor, Free and Slave: Workingmen and the Anti-slavery Movement in the United States.* New York: Associated Authors.

Marrant, John. 1785. *A Narrative of the Lord's Wonderful Dealings with John Marrant, A Black.* London: Gilbert and Plummer. Internet Archive.

Mazel, David. 2000. *American Literary Environmentalism.* Athens: University of Georgia Press.

McFeely, William S. 1991. *Frederick Douglass.* New York: Norton.

McGivikan, John R., and Stanley Harrold. 1999. *Antislavery Violence: Sectional, Racial, and Cultural Conflict in Antebellum America.* Knoxville: University of Tennessee Press.

McInerney, Daniel J. 1994. *The Fortunate Heirs of Freedom: Abolition and Republican Thought.* Lincoln: University of Nebraska Press.

Miller, Daegan. 2018. *This Radical Land: A Natural History of American Dissent.* Chicago: University of Chicago Press.

Montrie, Chad. 2008. *Making a Living: Work and Environment in the United States.* Chapel Hill: University of North Carolina Press.

Myers, Jeffrey. 2005. *Converging Stories; Race, Ecology, and Environmental Justice in American Literature.* Athens: University of Georgia Press.

National Park Service. 2016. *A Call to Action: Preparing for a Second Century of Stewardship and Engagement.* National Park Service. https://www.nps.gov/calltoaction/PDF/C2A_2015.pdf

———. n.d. Grand Canyon Hosts Camping 101 and the Camp Moreno Project. *Grand Canyon National Park.* https://www.nps.gov/grca/learn/news/grand-canyon-hosts-camping-101-and-the-camp-moreno-project.htm

———. n.d. Yosemite National Park Ranger Shelton Johnson wins National Freeman Tilden Award. *Yosemite National Park.* https://www.nps.gov/yose/learn/news/shelton-tilden.htm

Newman, Lance. 2009. Free Soil and the Abolitionist Forests of Frederick Douglass's 'The Heroic Slave'. *American Literature* 81 (1): 127–152.

OutdoorAfro. We Celebrate and Inspire African-American Connections to Nature. Accessed 5 Sept 2014.

Outka, Paul. 2008. *Race and Nature from Transcendentalism to the Harlem Renaissance.* New York: Palgrave Macmillan.

Peterson, Jodi. 2014. Parks for All? *High Country News* 46 (8): 11–18.

Pierson, Michael D. 2003. *Free Hearts and Free Homes: Gender and American Antislavery Politics.* Chapel Hill: University of North Carolina Press.

Purdy, Jedediah. 2016. Environmentalism Was Once a Social-Justice Movement. It Can Be Again. *The Atlantic*, December 7.

Quarles, Benjamin. 1969. *Black Abolitionists*. New York: Oxford University Press.

Reed, T.V. Towards an Environmental Justice Ecocriticism. In ed. Adamson, Evans, and Stein, 145–162.

Rice, Alan J., and Martin Crawford. 1999. Triumphant Exile: Frederick Douglass in Britain, 1845–1847. In *Liberating Sojourn: Frederick Douglass and Transatlantic Reform*, ed. Alan J. Rice and Martin Crawford, 1–12. Athens: University of Georgia Press.

Riss, Arthur. 2006. *Race, Slavery, and Liberalism in Nineteenth-Century American Literature*. New York: Cambridge University Press.

Roediger, David. 1991. *The Wages of Whiteness: Race and the Making of the American Working Class*. London: Verso.

Ronda, Margaret. 2012. 'Work and Wait Unwearying': Dunbar's Georgics. *PMLA* 127 (4): 863–878.

Ruffin, Kimberly N. 2010. *Black on Earth: African American Ecoliterary Traditions*. Athens: University of Georgia Press.

Sale, Maggie. 1992. Critiques from Within: Antebellum Projects of Resistance. *American Literature* 64: 695–718.

Sewell, Richard H. 1976. *Ballots for Freedom: Antislavery Politics in the United States, 1837–1860*. New York: Oxford University Press.

———. Slavery, Race, and the Free Soil Party, 1848–1854. In ed. Kraut 101–124.

Shore, Lawrence. 1986. *Southern Capitalists: The Ideological Leadership of an Elite, 1832–1885*. Chapel Hill: University of North Carolina Press.

Smith, Kimberly. 2007. *African American Environmental Thought: Foundations*. Lawrence: University Press of Kansas.

Spence, Mark David. 1999. *Dispossessing the Wilderness: Indian Removal and the Making of the National Parks*. New York: Oxford University Press.

Speth, James Gustave, and J. Phillip Thompson III. 2016. A Radical Alliance of Black and Green Could Save the World. *The Nation*. May 9–16.

Stauffer, John. Advent Among the Indians: The Revolutionary Ethos of Gerrit Smith, James McCune Smith, Frederick Douglass, and John Brown. In ed. McGivikan and Harrold, 236–273.

Taylor, Dorceta. 2014. *Toxic Communities: Environmental Racism, Industrial Pollution, and Residential Mobility*. New York: New York University Press.

Taylor, Patricia A., Burke D. Grandjean, and James H. Gramann. 2011. *Racial and Ethnic Diversity of National Park System Visitors and Non-Visitors*. In National Park Service Comprehensive Survey of the American Public, 2008–2009. National Park Service Social Science, July 2011.

United Church of Christ Commission for Racial Justice. 1987. *Toxic Wastes and Race in the United States*. Cleveland: United Church of Christ.

Walter, Krista. 2000. Trappings of Nationalism in Frederick Douglass's 'The Heroic Slave'. *African American Review* 34: 233–247.

West, Edward. 1848. Slavery at War with the Interests of the Free Laboring Classes. *The Liberator,* November 20. Accessible Archives.

Wilentz, Sean. 1984. *Chants Democratic: New York City and the Rise of the American Working Class, 1788–1850.* New York: Oxford University Press.

Wilson, Ivy G. 2006. On Native Ground: Transnationalism, Frederick Douglass, and 'The Heroic Slave'. *PMLA* 121: 453–468.

Yarborough, Richard. 1990. Race, Violence, and Manhood: The Masculine Ideal in Frederick Douglass's 'The Heroic Slave'. In *Frederick Douglass: New Literary and Historical Essays,* ed. Eric J. Sundquist, 166–188. Cambridge: Cambridge University Press.

CHAPTER 3

The Native Wilderness

LANDSCAPES OF INDIGENOUS ENVIRONMENTALISM

In the winter of 2013–2014, Gregg Deal (Paiute) toured U.S. cities performing an artwork called "The Last American Indian on Earth." He arrayed himself in mash-up costume consisting of an eagle-feather bonnet, fringed leggings, a hair-pipe breastplate, and a pair of Vans painted to look like beaded moccasins. As he rode the New York subways, paced the Lincoln Memorial, and visited the Smithsonian Museum of National History, he carried a cardboard sign that read, "This Used to Be Indian Land But Everything Went to Crap."[1] On his website, Deal explains that the "piece is a window into the funny, sarcastic, truthful, and even emotional journey of an artist using himself as an instrument of awareness" and that it "documents what happens when an unsuspecting public is confronted with the flesh-and-blood version of a stereotype."[2] This kind of strategic performance of identity has a long history among indigenous people in North America and around the world, particularly during struggles to regain or retain control of traditional territories and to protect them from development.

In the last four decades, the world has seen a surge of such campaigns. In the 1980s, Chico Mendes and the Rubber Tappers' Union fought to create indigenous extractive reserves in the rainforests of Brazil. In the 1990s, the playwright Ken Saro-Wiwa led the Movement for the Survival of the Ogoni People, which opposed Shell Oil's pollution of Ogoniland.

© The Author(s) 2019   91

L. Newman, *The Literary Heritage of the Environmental Justice Movement*, Literatures, Cultures, and the Environment, https://doi.org/10.1007/978-3-030-14572-9_3

For the past decade, Edwin Chota and the Ashéninka have sought legal title to Saweto, their homeland in the Peruvian Amazon, in order to prevent illegal loggers from deforesting the headwaters of the Alto Tamaya River. These three examples secured international media attention because Mendes, Saro-Wiwa, and Chota were all murdered by their opponents. However, hundreds, perhaps thousands, of similar campaigns have been organized around the world without garnering significant media coverage. And despite the lack of public attention, many of these campaigns have achieved important victories. On the Akwesasne Mohawk reservation that straddles the U.S.-Canadian border, Katsi Cook and the Mothers' Milk Project forced General Motors to dredge and dispose of polychlorinated biphenyls (PCBs) that the corporation had dumped in the St. Lawrence River. These industrial chemicals cause neurotoxicity and endocrine disruption, and they had accumulated in Mohawk women's bodies and passed to their breastfed children. The cleanup is ongoing. Meanwhile, Mohawk children continue to suffer serious health effects.[3] On the Navajo Nation in the Southwestern United States, groups like Diné Citizens Against Ruining Our Environment, Eastern Navajo Diné Against Uranium Mining, and the Navajo Nation Dependents of Uranium Workers Committee have successfully demanded remediation of tailings piles and compensation for radiation exposure victims. Still, hundreds of radioactive sites have not yet been addressed, and nearby residents continue to suffer elevated rates of various cancers. Similarly, in the Nevada desert, the Western Shoshone successfully opposed the U.S. government's plans to build a nuclear waste dump at Yucca Mountain. However, the health effects of more than 1000 aboveground nuclear tests on the Nevada Test Site continue to plague the tribe. In all of these cases, demands for the environmental protection or restoration of large tracts of land coincide with calls for the reinstatement of indigenous sovereignty on ancestral territory and for the recognition of traditional land use practices.[4]

Increasingly, indigenous organizations have explicitly stated the connections between their ecological and political concerns. For instance, the Indigenous Tar Sands Campaign in Western Canada connects the goals of protecting sensitive habitat from the oil industry and of ensuring the well-being of First Nations people: "Northern Alberta is ground zero with over 20 corporations operating in the tar sands sacrifice zone.... The cultural heritage, land, ecosystems and human health of First Nation communities ... are being sacrificed for oil money in what has been termed a 'slow industrial genocide.'"[5] Similarly, Idle No More, a broad-based Canadian First Nations campaign that was launched in 2012, called on the public to "join in a peaceful revolution, to honour Indigenous sovereignty, and to protect the land and water."[6] Idle No More grew instantly into one of the largest and most militant political movements in Canadian history. It organized "hundreds of teach-ins, rallies, and protests" in little more than a year. This remarkable success was driven not only by effective use of social media but also by the movement's broad political appeal: "Idle No More seeks to assert Indigenous inherent rights to sovereignty and reinstitute traditional laws and Nation to Nation Treaties by protecting the lands and waters from corporate destruction."[7] Patricia Gualinga, leader of the Sarayaku community in the Ecuadorian Amazon, which in 2012 won a landmark indigenous rights case in the Inter-American Court of Human Rights, puts this point crisply, "We can't feed our children oil."[8]

Idle No More and other recent indigenous movements show the limits of the "theory of post-materialism," which states that "rapid industrialization and urbanization lead both to a separation from nature and to a greater and self-conscious move to protect and identify with it."[9] According to this common line of reasoning, it is only the world's economic elite who can *afford* to care about nature. But in fact, the environmentalism of the poor has flourished around the world for decades, even centuries. And where mainstream environmentalism has often focused solely on preserving wilderness for recreational purposes, indigenous environmentalism has consistently connected ecological issues with "questions of human rights, ethnicity, and distributive justice."[10] However, these two traditions do not simply run in parallel. Indigenous environmentalism presents a direct intellectual and political challenge to mainstream activists for whom the preservation of pristine habitat trumps all other concerns. Because the "environmentalisms of the poor [often] originate in social conflicts over access to and control over natural resources," mainstream environmentalists have often ignored or even opposed these struggles, seeing indigenous claims as additional threats to already threatened wilderness.[11] For instance, in the late 1990s, the Sierra Club, Greenpeace, the Sea Shepherd Society, and others vigorously opposed the Makah reassertion of traditional whaling rights off the coast of Washington.[12] More recently, the Audubon Society, the Sitka Conservation Society, the Southeast Alaska Conservation Council, and others have opposed the Sealaska Corporation's bid for control of a section of the Tongass National Forest. Sealaska is one of 13 native-owned corporations created under the 1971 Alaska Native Claims Settlement Act, by which the federal government returned more than

200,000 square miles of land to indigenous control. Sealaska hopes to exercise the last of its land selection options to gain ownership of about 0.5% of the Tongass. The corporation states that it will engage in sustainable forest management in order to provide long-term jobs for many of its 20,000 Tlingit, Haida, and Tsimshian shareholders. Mainstream environmentalists oppose the plan because the land in question includes significant tracts of old-growth temperate rainforest that are considered critical habitat with the potential for wilderness designation.[13]

The Makah whaling and Sealaska logging controversies are just two recent examples in a long history of environmentalists "protecting" wilderness from indigenous people. Karl Jacoby explains that the rise of the conservation movement and the passage of large-scale conservation legislation in the late nineteenth and early twentieth centuries radically altered the relationship of indigenous people to the land in North America:

> The movement's arrival shut off vast portions of tribal hunting and foraging areas, while also inhibiting Native Americans use of fire to shape the landscape around them. Even more strikingly, conservation interlocked on multiple levels with other, ongoing efforts – treaties, the establishment of reservations, allotment – to displace Indians' claims on the natural world in order to open up such areas to non-Indians. In this sense, conservation was for Native Americans inextricably bound up with conquest – with a larger conflict over land and resources that predated conservation's rise.[14]

Some of the most revered wild places in the United States were created by acts of racial cleansing that made it possible for them to be viewed by white visitors as pristine and untrammeled. For instance, the U.S. Army removed large bands of Crows, Shoshones, and Bannocks from Yellowstone in the late nineteenth century. The Park Service excluded the Blackfeet from Glacier National Park. The Ahwahneechees were barred from the Yosemite Valley. Likewise, the Havasupais were driven off the South Rim at Grand Canyon and restricted to a tiny reservation in the bottom of a flood-prone side canyon.[15]

This history of the ethnic cleansing of nature was driven by specific ideological causes. As William Cronon argues, when we imagine the "sacred wilderness" as a pristine place where there is no evidence of human impact, we implicitly contrast it with its polar opposite, the degraded "civilization" occupied by "profane humanity." This habit of thought "tends to cast any use as *abuse*, and thereby denies us a middle ground in which responsible use and non-use might attain some kind of balanced, sustainable relationship."[16] Indigenous environmentalism, on the other hand, has grounded itself in the idea that human use of the land is fundamental to the well-being of both people and nature. The natural community is incomplete without active human membership.

As a result of this integrative conceptual framework, indigenous environmentalism frequently manifests itself in the form of campaigns for land ownership and management authority. Jace Weaver (Cherokee) explains that native sovereignty means not just tribal self-determination, but "power over the environment in which native peoples live [and power over] issues affecting their people's health, the air they breathe, the water they drink and from which they get their sustenance, the safety of the soil on which they walk and from which they derive their food."[17] When they lay claim to ancestral territory, indigenous people often argue that the traditional ecological knowledge they have accumulated over many generations allows them to manage the land more effectively than remote government agencies whose decisions are driven by Western science and politics. Winona LaDuke (Anishinaabeg) defines traditional ecological knowledge (often abbreviated TEK in legal and land management texts) as "the culturally and spiritually based way in which indigenous peoples relate to their ecosystems. This knowledge is founded upon spiritual-cultural instructions from 'time immemorial' and on generations of careful observation within an ecosystem of continuous residence."[18] In response to native land claims that invoke TEK, mainstream environmentalists have often argued that indigenous people cannot be trusted to act as responsible stewards of the land. For instance, in his 1999 book, *The Ecological Indian*, environmental anthropologist Shepherd Krech traces the centuries-long evolution of the contemporary image of "the Native North American as ecologist and conservationist" from its roots in the Euro-American trope of the noble savage.[19] He argues that this image is not only false but also harmful, because it masks the full human complexity of indigenous people's many different kinds of relationships with their different environments. In support of this claim, he details a number of examples, including the Pleistocene mega-fauna extinctions, the more recent overhunting of white-tailed deer and other game species, and the collapse of Hohokam irrigated agriculture due to soil salinization. He concludes that indigenous peoples have used natural resources no more sustainably than any other human group that must eke out a subsistence from the land.[20] Krech's work set off an intense controversy that continues today.[21] Many scholars have produced alternative historical examples of indigenous groups who stewarded their natural resources wisely. Others have attempted to show that indigenous intellectual and spiritual traditions are more ecologically aware than Euro-American philosophies and religions.[22] Perhaps the most insightful response comes from environmental historian Joy Porter, who agrees that indigenous people are no more inherently or authentically ecological than any others, but then argues that during the era of colonialism, they developed "strategies for coping with environmental damage, cultural engulfment, and forced migration" from which we have much to learn. "Indian history, literature, spirituality, and culture," Porter writes, "hold much that is too valuable to ignore, not least the native experience of what it means to live with profound environmental and spiritual loss."[23] According to this view, indigenous environmentalism is not a static inheritance from pristine pre-contact cultures, it is a set of active convictions that indigenous people have formed in response to seeing their land ravaged for profit.[24]

When indigenous people have successfully reestablished legal authority over traditional territories, they have often attempted to provide for sustainable use that preserves the long-term integrity of the entire biocultural community. In the late 1990s, for example, the InterTribal Sinkyone Wilderness Council, which represents ten tribes with ties to the Eel River area in northern California, acquired almost 4000 acres of wilderness from the lumber giant Georgia Pacific. The Council now manages this tract of land under the terms of two conservation easements that restrict commercial logging, grazing, occupancy, and other high-impact activities. The Council permits limited recreational access to a network of trails that link with trails on nearby public lands. The easements also provide for traditional uses such as harvesting trees for habitat management and traditional purposes, as well as fishing and gathering plants for medicinal and craft uses.[25] Similarly, in the last decade, the Native American Land Conservancy has worked to establish native-managed reserves at multiple locations on ancestral Cahuilla and Chemehuevi lands in Southern California, including the Old Woman Mountains Preserve, which the Conservancy manages according to the terms of an adaptive management plan that was developed with input from the U.S. Fish and Wildlife Service and nearby conservation organizations. The most prominent native conservation effort in the United States has centered on Bears Ears National Monument, where an intertribal coalition consisting of the Navajo Nation, Hopi, Ute Mountain Ute, Ute Indian Tribe of the Uintah and Ouray Reservation, and the Pueblo of Zuni partnered with the U.S. Forest Service and the Bureau of Land Management to manage 1.35 million acres in southwestern Utah. President Barack Obama's remarkable proclamation of the monument directly invokes indigenous environmentalism, noting that "the Navajo refer to such places as 'Nahodishgish,' or places to be left alone," and it creates a Bears Ears Commission consisting of delegates of the five tribes, who will participate in management decisions in perpetuity.[26] An accompanying presidential statement declares that the goals of management will be to "ensure that future generations are able to enjoy and appreciate these scenic and historic landscapes [and] that tribes and local communities can continue to access and benefit from these lands for generations to come."[27]

On an even larger scale, in 2011, the Taku River Tlingit first nation and the provincial government of British Columbia negotiated an agreement, the *Wooshtin Yan too.aat* (Walking Together) *Land and Resource Management and Shared Decision Making Agreement*, that provides for shared management of more than 15,000 square miles (almost 10 million acres) of boreal and temperate forest in the Taku River Tlingit traditional territory.[28] Working within the framework established by this agreement, representatives of the first nation and the provincial government negotiated the *Wóoshtin Wudidaa* (Flowing Together) *Land Use Plan*, which is "a comprehensive strategic land-use vision and plan addressing a wide range of natural and cultural resources, including the establishment of 11 new First Nations-focused conservancies, 13 Special Management Zones, over 180 landscape-level 'Tlingit Cultural Sites' and a 7 million acre 'Forest Retention Zone,' where commercial logging is not allowed."[29] The *Wóoshtin Wudidaa* is guided by a clear mission statement:

> Our common vision is for communities that are supportive, secure, and healthy, where people enjoy the peace and beauty of their natural surroundings and a sustainable quality of life.
>
> Ecosystems are healthy and fully functioning, and special wilderness areas and cultural places are protected. The natural environment is productive and supports diverse and abundant animal, fish and plant species as well as sustainable opportunities for harvesting, gathering and other activities on the land, including the Tlingit land-based way of life – *Hà khustìyxh* – and the lifestyle of the local community.
>
> Economic activity is diverse and vibrant, providing enduring employment and contributing to a just and prosperous future for our communities. Tlingit traditional land use has been sustained and revitalized, and exists in harmony with contemporary local land use....

> Collectively, we are living up to a shared responsibility to manage the land and resources in a way that honors our elders and ensures that we meet the needs of today without compromising opportunities for future generations.[30]

The Taku River Tlingit traditional territory is larger than the U.S. state of Maryland and, by almost any measure, it is wild. But it is a peopled wilderness. And the land use plan that governs it sets out to manage this working landscape so that it will continue to serve as healthy habitat for humans and all other beings.

The work of the Taku Tlingit and others shows vividly that, regardless of its origins or authenticity, the so-called myth of the ecological Indian powerfully informs and inspires indigenous environmentalism in the present.[31] This performative identity is very often underwritten by an equally powerful imaginative landscape, the native wilderness *topos*, which offers a retrospective vision of a native community making a peaceful home in pristine Nature. For instance, the *Wóoshtin Wudidaa Land Use Plan* includes a collective statement of "Tlingit Perspectives on Land and Resources":

> Our ancestors named the Taku River and still today we identify ourselves with this life-sustaining river as the Taku Quan, or people of the T'akhu. There is archeological evidence showing our people lived in the Taku watershed for at least the past 6,000 years.... Throughout all this time, our way of life, or *khustiyxh* has become intertwined with our lands and waters, so that we are now inseparable from these very same lands and waters.... This is the place we call home. This is our true homeland. Other people may come and go, but we have always been here and we will continue our lives here forever. We are part of the land, part of the water and part of the air in our territory – and without healthy water, land and air, we will no longer be who we are today.[32]

In this ringing passage, the Taku Tlingit assert that, as a people, they were created and are sustained in body, spirit, and mind by the land they live on. In a document that is otherwise quite utilitarian and bureaucratic, this paragraph stands out for its self-consciously literary language. The words *Taku*, *land*, and *water* are repeated rhythmically in short, declarative sentences that enact the rhythms of ceremonial speech. The passage has been made to contrast strongly with its surroundings by typographical means: it is set in a darker, more substantial font, and it is marked off by a thick, black border. Clearly, the creators of this document are making very deliberate choices as they evoke the native wilderness *topos* in order to establish their ecological credentials. After all, they need to convince both the provincial government and the people of British Columbia, including loggers, miners, conservationists, hunters, and fishing enthusiasts, that they not only have a right to manage this tract of land but also that they will steward it wisely and effectively. At the same time, this performative identity is more than just a rhetorical tool for establishing authority. It also informs the management principles stated in the plan, which in turn direct the actual practices now in effect on the land. The key goals of the *Wóoshtin Wudidaa Land Use Plan* are to:

1. Provide for cultural, social, and economic activities that support and balance healthy, resilient, and sustainable communities and economies and that generate lasting local and provincial benefits.
2. Sustain diverse and healthy native biodiversity, wildlife, fish, and ecosystems across the landscape in perpetuity.
3. Give special attention to ecosystems that are rare, or at risk, and species that are at risk or are of special management concern, so that biodiversity across the landscape is maintained.
4. Sustain and enable the continuation of Tlingit *khustìyxh* (way of life).[33]

The plan attempts to balance the needs of all inhabitants, human and non-human, of the Taku watershed. Of course, conflicts and problems will arise that cannot be resolved; nevertheless, the *Wóoshtin Yan Too.aat Shared Decision Making Agreement* and the *Wóoshtin Wudidaa Land Use Plan* lay out an astonishingly comprehensive framework for cultural and ecological sustainability and conservation-based decision-making.

The Taku Tlingit are participating in a long-standing tradition in which indigenous activists in North America have harnessed stereotypes, performed pan-Indian identities, and repurposed Euro-American landscape *topoi* in order to lay claim to ancestral lands and to make broader anti-colonial political claims. Artist and activist Russell Means (Lakota), who was a key leader of the American Indian Movement, writes, "I think we should be very happy and proud that we are still Indian people. We're still alive, and we're still resisting. We still have respect for the earth. We have traditional knowledge and values that are superior to anything in Western, 'scientific,' industrialized culture."[34] When Means accepts the term Indian, a colonial misnomer that collapses the thousands of North American indigenous tribal cultures into a single homogeneous identity, he does so because Indian is implicitly opposed to White and can be inhabited as a position of resistance. Winona LaDuke invokes the continuity of this tradition of pan-Indianism in an article that recounts the rise of the Indigenous Environmental Network and the work of its member organizations:

> Across the continent, on the shores of small tributaries, in the shadows of sacred mountains, on the vast expanse of the prairies, or in the safety of the woods, prayers are being repeated, as they have been for thousands of years, and common people with uncommon courage and the whispers of their ancestors in their ears continue their struggles to protect that land and water and trees on which their very existence is based. And like small tributaries joining together to form a mighty river, their force and power grows. This river will not be dammed.[35]

As LaDuke's moving evocation of the native wilderness *topos* makes clear, the performative tradition of indigenous environmentalism is vividly aware of its deep cultural roots.

All environmentalists today should be aware of this important thread of cultural history. A good place to begin is with two of the first Native American autobiographies written in English: *A Son of the Forest* (1829) by William Apess (Pequot) and *The Life, History, and Travels of Kah-ge-ga-gah-bowh* (1850) by George Copway (Anishinaabe). Lavonne Brown Ruoff compares these texts in literary and historical importance to the slave narratives of Olaudah Equiano, Frederick Douglass, and others. She also suggests that Native American and African American authors faced similar rhetorical challenges:

> On the one hand, the narrators had to convince their [mainly white] readers that they were members of the human race whose experiences were legitimate subjects of autobiography and whose accounts of these experiences were accurate. On the other hand, they had the moral obligation to portray the harsh injustices they and their fellows suffered at the hands of Christian whites.[36]

Both Apess and Copway address these competing demands by framing their autobiographies as Christian conversion narratives. Within that frame, they take on a difficult double task: they both praise the virtues of their tribes' traditional lifeways, and they pledge that their people will assimilate to white civilization after being educated in English language, culture, and religion.[37] Not surprisingly, these texts have not fared well with contemporary readers of Native American literature, who oppose cultural imperialism and tend to favor bold assertions of native sovereignty. As a result, Apess and Copway have received relatively little attention, not just from environmentalists but also from literary scholars.[38] Nevertheless, they have much to teach us about the rhetoric of indigenous environmentalism, since both authors fluidly perform pan-Indian identities while asserting the sovereignty of their tribes and laying claim to traditional homelands.

GEORGE COPWAY/KAHGEGAGAHBOWH AND IDENTITY AS PERFORMANCE

George Copway/Kahgegagahbowh was born into the Rice Lake Mississauga band in an active contact zone, the Trent River valley just north of Lake Ontario.[39] The Mississauga, a subgroup of the Anishinaabeg, had migrated to the area from the east a century earlier, and they now subsisted by hunting, trapping, fishing, and gathering. However, soon after Copway's birth, the Canadian government began settling recent European immigrants in the area. Open country was quickly being converted to farms, and along with settlers came disease, alcohol, and missionaries. This onslaught of material and cultural pressures destabilized the Rice Lake band, and their numbers dropped sharply as traditional lifeways became more and more unreliable. During this period of instability, Copway's parents converted to Methodism under the ministry of Peter Jones, a bicultural Mississauga/Canadian missionary who preached temperance and encouraged his converts to take up a settled life of agriculture. They sent their son to a mission school where he learned to speak, read, and write English. In 1830, when he was 12, he followed their example and converted to Christianity at a camp meeting. Four years later, he answered a call from the Methodist Church to serve a mission to the Anishinaabeg living on the shores of Lake Superior, several hundred miles to the west. For the rest of his life, Copway would act as an intermediary between the Anishinaabeg and their colonizers.

After completing his first mission, he spent two years as a seminary student at the Ebenezer Manual Labor School near Jacksonville, Illinois. Returning to Canada, he met Elizabeth Howell, and the two were married in 1840 despite intense white opposition to their interracial partnership. For the next few years, the couple conducted missionary work among the Anishinaabeg and Sioux across the northern Great Lakes region. In 1846, Copway was accused of misusing funds belonging to the Saugeen and Rice

Lake bands. He was expelled from his church and imprisoned briefly. When he was released, he moved to the United States, where in 1847 he published his autobiography, *The Life, History, and Travels of Kah-ge-ga-gah-bowh (George Copway)*. The book sold through seven printings in two years, and its author became an instant celebrity on the lecture circuit along the Atlantic seaboard. He took advantage of his newfound visibility to argue that the U.S. federal government should establish a sovereign territory for the indigenous people of the Great Lakes region—a proposal he detailed in *Organization of a New Indian Territory, East of the Missouri River* (1850). The tribes would govern themselves, build an agricultural economy, and eventually become U.S. citizens when the territory was converted into a new state.[40]

Copway was lionized by prominent American politicians and by literary figures like Washington Irving, Francis Parkman, James Fenimore Cooper, Henry Wadsworth Longfellow, and William Cullen Bryant. His proposal received significant political support and spurred intense public discussion of native sovereignty. Encouraged, he sought broader, international audiences for his ideas and writings. He brought out new editions of his autobiography in New York in 1850 and London in 1851, and he published a new book, *The Traditional History and Characteristic Sketches of the Ojibway Nation*, in London, Edinburgh, and Dublin in 1850 and in Boston in 1851. An epic poem, *The Ojibway Conquest: A Tale of the Northwest*, appeared in 1850 with Copway listed as the author on the title page.[41] During this period of hectic literary activity, Copway also traveled to Europe, where he addressed the third annual World Peace Congress in Frankfurt in August 1850. The German press celebrated his appearance enthusiastically. In order to capitalize on his now international fame, he rushed out an account of his European travels, *Running Sketches of Men and Places, in England, France, Germany, Belgium, and Scotland* (1851). He also launched *Copway's American Indian*, a New York newspaper that covered indigenous political affairs, but he was unable to attract enough financial support, and the paper folded after three months. By 1852, his prominent supporters as well as the general public had begun to lose interest. Though his *Traditional History* was reissued in 1858 and again in 1860, his brief career as a writer, editor, and lecturer was essentially over. For the next 20 years, he struggled to support his family, and he relocated them frequently. He unsuccessfully sought backers for new speaking tours, petitioned for employment as a federal reservation agent, worked for a time as a U.S. Army recruiter in Canada, and later practiced herbal medicine in Detroit. He died in 1869 while working as a mediator between Catholic missionaries and an Iroquois band at the confluence of the Ottawa and St. Lawrence Rivers near Montreal.

Copway translated his Mississauga name, Kahgegagahbowh, as "Standing Firm." However, he reinvented his public identity repeatedly throughout his life. He began as a traditional Mississauga youth, then transformed himself into a Canadian Methodist Indian missionary, then into an American patriot, and then into a refined literary globe-trotter. In all of these roles, he displayed a keen understanding of his audiences, and he adapted his rhetoric and persona to their expectations in ways that were designed to make his ideas intelligible and persuasive. For instance, during his lectures proposing a new Indian territory, he routinely appealed for support to the anti-immigrant ultra-patriots of the Know-Nothing Movement by identifying himself as a true "native" American who shared his listeners' love of their shared birthright. During his appearance at the World Peace Congress, he was billed as "a chief of the Red Indian tribes." On stage, he wore metallic armbands over his suit, carried a decorated staff, and gave the president of the Congress an ornate calumet, which he described as "a weapon of peace." In his address, he struck a pose of mock humility, protesting that he had only recently learned "the language of his pale brothers." Moreover, he told the assembly, "I am the first of my race to journey all the way from America's wilderness in order to assist in establishing peace over here." Finally, he took the opportunity to move a resolution that "That this Congress, acknowledging the principle of non-intervention, recognizes it to be the sole right of every state to regulate its own affairs."[42] With remarkable savvy, Copway performed a stock "noble savage" identity that warranted his political argument by engaging the delegates' Romantic beliefs about the natural morality and generosity of "primitive" people.[43]

Copway's writings are just as fluid and performative as his public selves. His authorial personas and his rhetorical appeals shift repeatedly to suit new situations and purposes, sometimes on multiple occasions within the pages of a single book. Not surprisingly, literary historians have focused mainly on his authorial identity and have struggled to explain its instability. Some have been dismayed by what they see as his inauthenticity and even dishonesty. For instance, Donald Smith, Copway's principal biographer, warns readers to "Beware! – as even a cursory investigation of his past reveals that historians and anthropologists must use his work with great caution." He even compares Copway to "an aggressive American promoter" and claims that he routinely "twisted facts to place himself in the best possible light."[44] Rather than view him as an impostor, Cheryl Walker argues that Copway "doesn't seem to quite know who he is" and even that he suffers from a "lack of psychological and cultural coherence."[45] Joshua Bellin, on the other hand, warns us that "to dismiss Copway as a 'spurious' Indian based on his apparent distance from 'traditional' standards is to assume that stories, cultures, [and] traditions exist in pure ... form rather than as local acts in contexts of encounter."[46] Following Bellin's lead, most recent commentators reject the idea that Copway was a docile subject of ideological indoctrination, and they explain his fluid identity as a strategy of survivance in a contact zone. Cathy Rex, for instance, celebrates his creation of "a new possibility for Indianness, a self-determined identity that defies and resists ... static national, racial, social, and intellectual categories."[47] Still others admire him as a latter-day Trickster who flamboyantly and sometimes brilliantly manipulated both Romantic stereotypes and the details of Anishinaabe culture and history in order to awaken sympathy in American and European audiences.[48] After all, Copway faced the desperate task of convincing racist Euro-Americans to support his demand that the U.S. government abandon its official policy of genocide and establish a permanent native homeland. It is no wonder then that he switched masks so often. He was working with all of the rhetorical tools that came to hand. When one failed, he threw it away and picked up a new one without a second thought.

AUTHORIAL IDENTITY IN COPWAY'S *LIFE, HISTORY, AND TRAVELS*

The first edition of Copway's autobiography was published in 1847 by Weed and Parsons in Albany, New York. For the frontispiece, the company chose an engraving from a watercolor by Felix Octavius Carr Darley, a prominent New York artist who had illustrated several of James Fenimore Cooper's Leatherstocking novels. Darley's image, which readers would inevitably view as a portrait of Copway, features a stereotypical noble savage standing on a hillock in front of a soft-focus forest.[49] He strikes a relaxed *contrapposto* pose with his arms crossed on a strung bow, and he is framed by a landscape of sculptural clouds, boulders, and tall pines that simultaneously emphasizes his nobility and displaces him into the native wilderness beyond the frontier. A caption at the bottom of the page reads, "I am one of Nature's Children." This sentence is a quotation from the first chapter of the book, where it occurs in the middle of a passage in which Copway lays claim to a pan-Indian identity.

In the second and later editions of the book, Darley's sublime image is replaced with a portrait of the author that emphasizes his status as a cultural and religious convert. In this dark engraving from a charcoal sketch, Copway's shoulders are set in three-quarter profile, but his face looks directly at the viewer with a direct and impassive stare.[50] His posture is upright, and he wears a simple, dark suit. His hair is trimmed just below the ears and combed to the side, revealing a high forehead, straight brow, and strong cheekbones. In the visual discourse of the period, the portrait emphasizes Copway's intelligence and moral vigor. However, it also presents him as an entirely Christianized and Europeanized Indian. By virtue of high contrast and side light, his skin color appears nearly white. He passively inhabits a blank interior space from which Nature has been absolutely excluded.

The book's front matter also projects the author's double identity, along with his alternately defiant and conciliatory attitude toward the colonizing culture to which most of his readers belonged. The full title directly juxtaposes his Mississauga and English names and then locates him in a sequence of roles: *The Life, History, and Travels of Kah-ge-ga-gah-bowh (George Copway), a Young Indian Chief of the Ojebwa Nation, a Convert to the Christian Faith, and a Missionary to His People for Twelve Years*. The main phrase here boldly suggests that the author's indigenous identity remains primary, while his colonized self is merely parenthetical. However, the subordinate phrases that follow, which identify him as both a chief and a convert, maintain that he is perfectly suited to play the intermediary roll of missionary to his people. In order to dispel the white reader's worries about the genuineness of Copway's shift of cultural allegiance, the title page attests to his literacy by stating prominently that the book was "Written by Himself."[51] The dedication page even more forcefully retracts Kahgegagahbowh's unsettling prominence in the title by adopting the deferential stance of a loyal colonial subject and by infantilizing natives in general. It describes the book as a "Brief History of a Child of the Forest" and presents it to the "Clergy and Laity of the American and British Dominions."[52]

In "A Word to the Reader," Copway addresses the issue of authorship at greater length, and in doing so, he again alternates between defiant indigeneity and the deference of the convert. He begins by making a roundabout apology for his language skills: "It would be presumptuous in one, who has but recently been brought out of a wild and savage state; and who has since received but three years' schooling, to undertake, without any assistance, to publish to the world a work of any kind. It is but a few years since I began to speak the English language." In this passage, he seems to accept the colonial racial hierarchy that defines him as inferior, especially when he states that he is incapable of writing without help. At the same time, though, by situating himself in an implied narrative about the acquisition of English and "schooling," he insists on his own capacity to cross linguistic and cultural borders. In the following sentence, he humbly states his gratitude for the assistance of "a friend, who has kindly corrected all serious grammatical errors." Finally, he retracts his ambivalent apology altogether by insisting that "the language except in a few short sentences, the plan, and the arrangement [of the book] are all my own."[53] Of course, the highly polished and conventional discourse of this passage belies his claim of linguistic and cultural ignorance, so in the very act of asking for his readers' indulgence, he establishes his own legitimacy and distinctiveness as a Native American author.

Copway closes his prefatory note with a rhetorical flourish in which he again transforms abjection into a position of strength. In a theatrical apostrophe, he laments his own dislocation from home as a way of establishing a rapport with the reader that is rooted in sympathy:

> I am a stranger in a strange land! And often, when the sun is sinking in the western sky, I think of my former home; my heart yearns for the loved of other days, and tears flow like the summer rain. How the heart of the wanderer and pilgrim, after long years of absence, beats, and his eyes fill, as he catches a glance at the hills of his nativity, and reflects upon the time when he pressed the lips of a Mother, or Sister, now cold in death.[54]

The mournful tone here suggests that he has resigned himself to permanent exile, perhaps even to being the last of his tribe. At the same time, by quoting *Exodus*, he suggests that he will one day return to redeem his people from oppression. In this biblical story, the young Moses kills an Egyptian to defend a fellow Hebrew, then flees to the neighboring "land of Midian." There, he marries Zipporah, who "bare him a son, and he called his name Gershom: for he said, I have been a stranger in a strange land." Moses remains in Midian until God sends him back to Egypt to lead the "children of Israel" to "a land flowing with milk and honey."[55] As if to compensate for subversively comparing himself to Moses, Copway concludes the preface on a submissive note. He begs the "friends of humanity" to "direct their benevolence to those who were once the lords of the land on which the white man lives."[56] This gesture of obeisance is so hyperbolic that it can only be read as flattery:

> Pray for us – that religion and science may lead us on to intelligence and virtue; that we may imitate the good white man, who, like the eagle, builds its nest on the top of some high rock – science; that we may educate our children, and turn their minds to God. Help us, O help us to live – and teach us to die a Christian's death, that our spirits may mingle with the blessed above.[57]

Just after implicitly representing himself as a savior who will lead God's chosen people out of bondage, he puts on the mask of the Christian Indian and sends the white/Egyptian reader off into the story of his conversion with a wink and a pat on the back.

Copway's two personas—the Christian Indian and the defiant Mississauga—coexist throughout *The Life, History, and Travels*. The first persona is dominant, since the book is framed as a conversion narrative that uses the author's experience to demonstrate Native American potential for cultural assimilation. However, the second emerges regularly to protest the cultural disintegration that occurs when indigenous people are displaced from their traditional lands. The productive tension between these two personas is established at the beginning of the first chapter: "I loved the woods, and the chase. I had the nature for it, and gloried in nothing else. The mind for letters was in me, but was asleep, till the dawn of Christianity arose, and awoke the slumbers of the soul into energy and action."[58] These are the first sentences in which Copway describes his own experience, and they directly juxtapose a traditional indigenous life of the body on the land with the Christian life of the mind in letters and of the soul in God. While he claims to have renounced "the chase," he does so ambivalently at best.

Copway's intense attachment to traditional Mississauga lifeways shows clearly throughout the book's first half, which ostensibly disparages the "wild and savage state" in which he existed before adopting Methodism. However, rather than depict himself and the Rice Lake band as backward and ignorant, he enthusiastically celebrates traditional Mississauga culture. For instance, when he describes the pantheon of "*mon-e-doos*" or spirits that he learned about as a child, his tone is worshipful: "At early dawn I watched the rising of the *palace* of the Great Spirit – the sun – who it was said, made the world!"[59] His stated purpose is to illustrate the depth of his religious slumber in his youth, but the warmth of his prose has the opposite effect. He particularly emphasizes the immanence of these tutelary spirits in Nature:

> My father taught [me that it was] *Ke-sha-mon-e-doo* – Benevolent spirit [who] made the earth, with all its variety and smiling beauty. His benevolence I saw in the running of the streams, for the animals to quench their thirst and the fishes to live; the fruit of the earth teemed wherever I looked. Every thing I saw smilingly said Ke-sha-mon-e-doo nin-ge-oo-she-ig – the Benevolent spirit made me.[60]

By providing and then translating Mississauga names and phrases in this passage, he insists on the specificity and integrity of tribal spiritual traditions. And by acknowledging his tribe's traditions so reverently at the beginning of his autobiography, he implies that they deserve to be maintained.

The defiant Mississauga steps forward again when Copway provides the information about his parentage that is demanded by the conventional structure of an autobiography. He converts this obligatory rehearsal into an opportunity to condemn European disregard of indigenous land claims. He begins by locating his parents in space: "My parents were of the Ojebwa nation, who lived on the lake back of Cobourg, on the shores of Lake Ontario, Canada West." The sequence of identifiers in this sentence first establishes his parents' cultural identity at the broadest level; they are Ojibwes, that is, a subgroup of the Anishinaabeg who migrated from the east a century earlier. Copway further refines their identity by placing them alongside Rice Lake where their band settled after arriving in the area.[61] Then he acknowledges the threat to their multilayered place-based identity that is presented by European renaming of the area as "Cobourg" in "Canada West." In his next sentence, he makes clear that this act of erasure is not merely linguistic; European settlement has directly impacted the quality of his parents' lives on Rice Lake, where "there was a great quantity of wild rice, and much game of different kinds, before the whites cleared away the woods."[62]

THE NATIVE WILDERNESS *TOPOS* IN *THE LIFE, HISTORY, AND TRAVELS*

When Copway speaks as Kahgegagahbowh, the scenery behind him often shifts. In place of the Canadian settlements, the native wilderness *topos* comes into view. By emphasizing the fertility and orderly management of the land when it was under native control, Copway establishes a benchmark to measure the ecosocial damage of colonization. For instance, as he recites his family history, Copway locates their home "at the head of Crow River, a branch of the River Trent, north of the Prince Edward District, Canada West."[63] He shows that their historical ties to this place precede the current place names by narrating the Mississauga conquest of the region, when "the Ojebwa nation defeated the Hurons, who once inhabited all the lakes in Western Canada." He proudly notes his own family's role in that history by remarking that his "great-grandfather was the first to settle at Rice Lake." This Mississauga pioneer chose the area because he was "a great hunter" and there was "an abundance of game of every kind."[64] Because he "had a crane for his totem," the "Crane tribe became the sole proprietors of this part of the Ojebwa land."[65] Copway explains that Anishinaabe customs of land tenure blend collective sovereignty with private access. While the tribe as a whole claims and inhabits the region, individual families exercise exclusive use rights in well-defined places: "The Ojebwas each claimed, and claim to this day, hunting grounds, rivers, lakes, and whole districts of country. No one hunted on each other's ground."[66] Long-standing "law and custom" provided for the enforcement of individual rights of access:

> If any person was found trespassing on the ground of another, all his things were taken from him, except a hand full of shot, powder sufficient to serve him in going straight home, a gun, a tomahawk, and a knife; all the fur, and other things, were taken from him. If he were found a second time trespassing, all his things were taken away from him, except food sufficient to subsist on while going home. And should he still come a third time to trespass on the same, or another man's hunting grounds, his nation, or tribe, are then informed of it, who take up his case. If still he disobey, he is banished from his tribe.[67]

This escalating sequence of punishments culminates in the trespasser's exile from the home ground in which individual and tribal identities are rooted. Copway implies that, in sharp contrast with the colonial frontier, the native wilderness rarely saw trespassing or poaching because it was so thoughtfully managed.

The native wilderness *topos* appears again when Copway describes his childhood home in a declamatory set-piece:

> I was born in nature's wide domain! The trees were all that sheltered my infant limbs – the blue heavens all that covered me. I am one of nature's children; I have always admired her; she shall be my glory; her features – her

robes, and the wreath about her brow – the seasons – her stately oaks, and the evergreen – her hair – ringlets over the earth, all contribute to my enduring love of her; and wherever I see her, emotions of pleasure roll in my breast, and swell and burst like waves on the shores of the ocean, in prayer and praise to Him, who has placed me in her hand. It is thought great to be born in palaces, surrounded with wealth – but to be born in nature's wide domain is greater still.[68]

In the context of the conversion narrative, the function of this passage is to show that God caused Copway to be born in primitive obscurity in order to prepare him for his role as a Christian missionary. However, God remains a distant and silent presence here, and the effusive language and breathless syntax make perfectly clear that Nature provokes much stronger feelings of devotion. Then, as if dissatisfied with the vague and highly conventional description he has provided so far, he particularizes his portrait of his home in the very next paragraph: "I was born sometime in the fall of 1818, near the mouth of the river Trent, called in our language, Sah-ge-dah-we-ge-wah-noong.... I remember the tall trees, and the dark woods – the swamp just by, where the little wren sang so melodiously after the going down of the sun in the west."[69] We move from the Trent, a generically named river in the Euro-American Romantic landscape, to Sah-ge-dah-we-ge-wah-noong in the homeland of the Mississauga. By giving his birthplace its proper name and by reciting a series of particular sensory images that show the depth of his bodily knowledge, Copway lays claim to his endangered inheritance. After doing so, he returns to the conventional language of Romanticism, but he now uses it to make a defiant prediction of cultural survival and even resurgence:

I would much more glory in this birth-place, with the broad canopy of heaven above me, and the giant arms of the forest trees for my shelter, than to be born in palaces of marble, studded with pillars of gold! Nature will be nature still, while palaces shall decay and fall in ruins. Yes, Niagara will be Niagara a thousand years hence! the rainbow, a wreath over her brow, shall continue as long as the sun, and the flowing of the river! While the work of art, however impregnable, shall in atoms fall.[70]

The Mississauga children of Nature will thrive in the native wilderness, while the invading Euro-American civilization will crumble to dust with all its vaunted cultural achievements and claims of racial superiority.

Having demonstrating that individual and tribal identities are tightly connected to specific material/symbolic environments, Copway now details how those environments have been degraded. The native wilderness was once extremely productive and provided the Mississauga with "large quantities of beaver, otter, minks, lynx, fishes, &c." But colonization has linked indigenous lands to faraway markets with predictable effects: "From these lakes and rivers come the best furs that are caught in Western Canada.... They are then shipped to New-York city, or to England.... Before the whites came amongst us, the skins of these animals served for clothing; they are now sold from three to eight dollars a piece."[71] The cheapening of indigenous lifeways by the operations of the fur market moves Copway to address the reader directly and condemn the sale of tribal territory:

> In the year 1818, 1,800,000 acres of [Mississauga land] were surrendered to the British government. For how much, do you ask? For $2,960 per annum! What a great sum for British generosity! Much of the back country still remains unsold, and I hope the scales will be removed from the eyes of my poor countrymen, that they may see the robberies perpetrated upon them, before they surrender another foot of territory.[72]

By now, we have traveled a considerable distance from the theme stated at the beginning of the book: the "unfortunate race called the Indians ... can be made to enjoy the blessings of [Christian] life."[73] For most of the book's first chapter, Copway has dropped the thread of the conversion narrative in order to extol the traditional Mississauga way of life in the native wilderness and to condemn the disruptive effects of colonization.[74]

In the second chapter, Copway details Mississauga methods of material subsistence prior to contact. His stated purpose is to illustrate the benefits of conversion by recalling "the actual condition of our people, before Christianity was introduced among us" and to thank God "for his preserving kindness to us, in sparing us to hear his blessed word."[75] And he begins by describing the intense bodily rigors of traditional life: "Some winters we suffered most severely, on account of the depth of snow, and the cold; our wigwams were often buried in snow. We not only suffered from the snow and the cold, but from hunger. Our party would be unable to hunt, and being far from the white settlements, we were often in want of food."[76] Nevertheless, Copway's pride in Mississauga ingenuity and toughness shows clearly when he describes how they build wigwams from animal skins, use birch bark canoes, make heavy portages, cover long distances on foot, hunt deer by sound at night, make cold-weather clothing out of bear skins, and gather wild rice that sustains them through long, hard winters. He recounts his father's campfire lectures on the ethics of hunting and lovingly describes how he learned to track, kill, and process game: "For years I followed my father, observed how he approached the deer, the manner of getting it upon his shoulders to carry it home. The appearance of the sky, the sound of distant water-falls in the morning, the appearance of the clouds and the winds, were to be noticed. The step, and the gesture, in traveling in search of the deer, were to be observed."[77] This chapter also features multiple scenes in which Mississaugas survive deadly weather because of their detailed traditional ecological knowledge. In one key anecdote, his father leads a hunting party that gets snowed in for several weeks. As they begin to starve to death, he relies on traditional divination and on his understanding of the habits of game animals to kill two beaver whose bodies sustain the group until the snow recedes enough to allow travel. As usual, the framing commentary on this episode cuts against its actual effect. Copway states that it illustrates the privation that the Mississaugas suffered prior to conversion, as well as God's providence in preserving them to become civilized. However, the story turns on his father's superior woodcraft, which saves the group from near death. Copway's pride and admiration for his tribe's traditional lifeways comes right to the surface when he nostalgically recalls his own prowess as a hunter: "I delighted in running after the deer, in order to head and shoot them. It was a well known fact that I ranked high among the hunters. I remember the first deer I ever shot, it was about one mile north of the village of Keene."[78] After presenting this detailed and passionate account of Mississauga material culture, Copway off-handedly reassures the Christian reader that he has left behind the life of the body on the land: "I loved to hunt the bear, the beaver, and the deer; but now, the occupation has no charms for me. I will now take the goose quill for my *bow* and its point for my *arrow*."[79] We are suddenly asked to believe that his attachment to Mississauga culture and the native wilderness remains safely in the past: "In the days of our ignorance we used to dance around the fire. I shudder when I think of those days of our darkness.... I thank God that those days will never return."[80] This pronouncement rings hollow.

ALCOHOL, METHODISM, AND THE SLOW VIOLENCE OF COLONIALISM

Copway returns to the subject of his tribe's cultural traditions in order to set the stage for condemning the European introduction of alcohol into indigenous bodies. Again, rather than represent traditional Mississaugas as primitive, he warmly describes their annual religious ceremonies, praises the ethical wisdom of the tribe's elders, and gives several examples of their sayings, such as "Never pass by any indigent person without giving him something to eat. Owh wah-yahbak-mek ke-gah-shah-wa-ne-mig – the spirit that sees you will bless you."[81] When such traditional principles held sway, he writes, "the lives, peace, and happiness, of the Indian race were secured; for then there was no whiskey amongst them."[82] What follows this blunt and sudden statement is not a temperance lecture directed at natives, but a political analysis of alcohol as a weapon of empire:

> O! that accursed thing. O! why did the white man give it to my poor fathers? … I recollect the day when my people in Canada were both numerous and happy; and since then, to my sorrow, they have faded away like frost before the heat of the sun! … The Ojebwa nation, that unconquered nation, has fallen a prey to the withering influence of intemperance…. They are hedged in, bound, and maltreated, by both the American and the British Governments.[83]

His tribe, which has never suffered a military defeat, has now been laid low by slow chemical violence.

It should come as no surprise, then, that when he narrates his father's conversion to Methodism, Copway does not describe it as a rejection of his tribe's traditional culture but presents it as a victory over the whites' poison. Several Mississaugas are drinking and, just as "the whiskey began to steal away their brains," a group of Indian missionaries appears.[84] After earnest conversation and prayer, his father "took the keg of whiskey, stepped into one of the small canoes, and paddled some thirty feet from the shore; here he poured out the whiskey into the lake, and threw the keg away."[85] He then convinces the rest of the group to accompany the missionaries to a camp meeting where they imbibe "*heavenly fire*" instead of "*fire-water*, or *devil's spittle*."[86] In this anecdote, Methodism serves not as a replacement for Mississauga *mon-e-doos* but as a means of redemption from the bodily pollution of colonialism. Moreover, it only proves the "enormous turpitude and recklessness" of "merciless, heartless, and wicked white men" that they have withheld Christianity from the Mississauga for so long while providing them with whiskey instead: "The least that Christians could have done, was to send the gospel among them, after having dispossessed them of their lands; thus preparing them for usefulness here, and happiness hereafter."[87] When he puts down the mask of the Christian Indian and speaks in the voice of the defiant Mississauga, Copway turns Christian ethical injunctions against the process of colonization, which he represents as a form of religious hypocrisy. The Mississaugas, he maintains, were generous, honest, and pious when governed by their own traditions, and now they are better Christians than the "pseudo Christian nation" from whom they have learned to worship Christ.[88]

The Life, History, and Travels turns on the moment of Copway's decision to become a Methodist missionary and "be useful to [his] brethren."[89] While the first half of the book takes place entirely in the native wilderness around Rice Lake, the second details his many long and difficult journeys to remote locations across the Great Lakes region where he works to convert Mississaugas and other Anishinaabeg. This second section can seem to be little more than a long list of departures, storms, portages, ambushes, episodes of hunger, miraculous deliverances, and late arrivals. However, the emphasis throughout is on the bodily stamina and strength that results from temperance and that also demonstrates the exceptional dedication of Copway and other Indian missionaries to using Methodism to help the people they serve withstand the slow violence of colonization. The section culminates with an account of a "General Council" of Christianized Anishinaabe near Lake Huron.[90] The Council has gathered to petition the Canadian government to set aside a substantial tract of traditional Saugeen Mississauga territory near Lake Huron for exclusive indigenous use. The Council members also discuss a plan for scattered and embattled Anishinaabe bands to "live together, and become a happy people, so that our dying fires may not go out [our nation may not become extinct], but may be kindled in one place, which will prove a blessing to our children."[91] Copway praises the Council for setting out "to adopt measures by which peace, harmony, and love, might be secured," and he implies that they are able to take collective action to secure permanent title to a new homeland because they have driven out whiskey with Methodism: "As I sat and looked at them, I contrasted their former (degraded), with their present (elevated) condition. The Gospel, I thought, had done all this."[92] This account of the General Council concludes with an indigenous political manifesto that once again condemns colonization as Christian hypocrisy:

With all the wholesome and enlightened laws; with all the advantages and privileges of the glorious Gospel, that shines so richly and brightly all around the white man; the poor ignorant Indians are compelled, at the point of the bayonet, to forsake the sepulchres of those most dear to them, and to retire to a strange land, where there is no inhabitant to welcome them!!! May the day soon dawn, when Justice will take her seat upon the throne.[93]

Copway explains that "the Indians are diminishing in numbers in the midst of their white neighbors" because of the "introduction of King Alcohol among them," because of "new diseases, produced by their intercourse with the whites," and because they are unable "to pursue that course of living, after abandoning their wigwams, which tends to health and old age."[94] In other words, colonization has polluted their bodies with alcohol and microbes, and it has alienated them from the land on which their traditional modes of healthy subsistence depended. He then asks white readers to imagine their own reaction to similar circumstances:

Our people have been driven from their homes, and have been cajoled out of the few sacred spots where the bones of their ancestors and children lie; and where they themselves expected to lie, when released from the trials and troubles of life. Were it possible to reverse the order of things, by placing the whites in the same condition, how long would it be endured? There is not a white man, who deserves the name of *man*, that would not rather die, than be deprived of his home, and driven from the graves of his relatives.[95]

The implications of this passage are clear. The Mississaugas would be perfectly well justified in responding to dispossession with violence. But instead of devising "schemes of murder," they have decided to take the path of righteousness and respond to oppression with faith and reasoned discourse: "Give us but the Bible and the influence of a Press, and we ask no more."[96] Using these very tools, Copway makes a series of clear and radical demands on behalf of "the Aborigines of America.[97]" He calls for "missions and high schools," equality before the law, political enfranchisement, and a clearly defined sovereign territory "so that they may represent their own nation."[98] These demands, which amount to a program for cultural survivance and tribal sovereignty within the framework of the U.S. nation-state, are directly supported by the doubled persona that Copway has maintained throughout the preceding narrative. Because *The Life, History, and Travels of Kah-ge-ga-gah-bowh (George Copway)* functions simultaneously as a conversion narrative and a narrative of awakening to political consciousness and commitment, it lends solid weight to the idea that, despite the dominance of the colonizers, indigenous peoples remain adaptable and resourceful inhabitants of their ancestral homes, where they are determined to do much more than survive. They will fight across gen erations for the right to access to their land, and they will work to maintain the well-being of the ecological systems that nourish their bodies and communities.

WILLIAM APESS, RACISM, AND INDIGENOUS IDENTITY

William Apess (1798–1839) was the most prolific of the Native American authors who were active during Andrew Jackson's presidency.[99] Apess (sometimes spelled Apes) suffered through a harsh childhood in a Massachusetts Pequot community that was ravaged by alcoholism and poverty. After the age of five, he was raised by maternal grandparents who physically abused him until he was removed from their care and indentured to a neighboring white family. At 11, after he was discovered to be planning an escape, he was sold to a second and then to a third white family, where he began attending Methodist camp meetings. At 15, he ran away, joined the U.S. Army, and fought in the War of 1812. When he mustered out, the Federal Government reneged on its promise to pay him in land and cash, and he worked his way around Eastern Canada for two years, before deciding to return to the United States. In 1821, he married Mary Wood and began working as an itinerant lay preacher to feed a quickly growing family. He was ordained in the Methodist Church in 1829 and continued to travel, preaching at camp meetings to mainly Native American and African American audiences. Apess settled for a time in the native town of Mashpee near the neck of Cape Cod, where a multi-racial community had formed that included escaped slaves, free blacks, poor whites, and others. He became a central leader of an 1833 campaign to convince the state of Massachusetts to recognize the Mashpees' sovereignty and to withdraw a minister who had been appointed by Harvard College and lived at the tribe's expense. During this campaign, Apess was jailed for 30 days for incitement to riot after forcibly unloading a white trespasser's cart of stolen firewood. Nevertheless, partly because of his powerful rhetorical skills, the "Mashpee Revolt" succeeded in winning limited tribal self-government and religious liberty. Apess died in New York City at the age of 41 in 1839.[100]

Despite the intense demands of his short and busy life, Apess published four books, seeking to use the power of the press to condemn racism and advocate for native rights:

> My people have no press to record their sufferings, or to make known their grievances; on this account many a tale of blood and wo, has never been known to the public. And during the wars between the natives and the whites, the latter could, through the medium of the newspaper press, circulate extensively every exaggerated account of "Indian cruelty," while the poor natives had no means of gaining the public ear.[101]

The text for which Apess is known today, "An Indian's Looking-Glass for the White Man," is a section of his second book, *The Experiences of Five Christian Indians of the Pequot Tribe* (1833). This collection of five conversion stories was designed to disprove the racist assumptions of white superiority that were common in the churches of New England at the time. The first edition of what is otherwise a quite discreet book contains a fiery concluding essay, "An Indian's Looking-Glass for the White Man":

> I would ask why are not [Native Americans] protected in our persons and property throughout the Union? Is it not because there reigns in the breast of many who are leaders, a most unrighteous, unbecoming and impure black principle, and as corrupt and unholy as it can be – while these very same unfeeling, self-esteemed characters pretend to take the skin as a pretext to keep us from our unalienable and lawful rights? I would ask you if you would like to be disfranchised from all your rights, merely because your skin is white, and for no other crime…. And now let me exhort you to do away that principle, as it appears ten times worse in the sight of God and candid men, than skins of color – more disgraceful than all the skins that Jehovah ever made. If black or red skins, or any other skin of color is disgraceful to God, it appears that he has disgraced himself a great deal – for he has made fifteen colored people to one white, and placed them here upon this earth.[102]

This passage remains one of the most direct and uncompromising condemnations of racism written in the nineteenth century.[103] At the essay's conclusion, Apess looks forward to the day when "this tree of distinction shall be leveled to the earth, and the mantle of prejudice torn from every American heart [for] then shall peace pervade the Union."[104] "An Indian's Looking-Glass for the White Man" has become a literary standard since it was republished in a 1982 anthology, *The Elders Wrote*.[105] Apess's other works deserve our attention too, since they record the extraordinary resourcefulness of a Native American writer who took on the daunting task of denouncing racism during an era of official genocide, while also calling for the restoration of indigenous sovereignty on the land.[106]

In order to appreciate the power of Apess's work, we need to think carefully about his frequent use of the discourses of Christianity and republicanism, which some readers have perceived as cultural treason. For instance, soon after "An Indian's Looking-Glass for the White Man" was republished, the influential scholar of Native American literature Arnold Krupat dismissed Apess as a colonial subject who had cravenly adopted the language, ideology, and identity of his oppressors: "Apes proclaims a sense of self, if we may call it that, deriving entirely from Christian culture."[107] Since then, other scholars have rejected the idea that by speaking the language of Christianity, Apess betrayed his Pequot heritage, and they suggest that by repurposing the discourses of colonialism, he was inventing rhetorical strategies of "survivance."[108] Rather than measure Apess's identity against an imaginary standard of indigenous authenticity, we need to recognize that his writings are savvy performances that repurposed colonial rhetorical strategies in order to condemn Jacksonian Indian policy, protest conditions on New England reservations, and lay the foundations for a radical pan-Indian political consciousness. So, in "An Indian's Looking-Glass," when Apess reproduces the rhythms of a fiery preacher who is trying to break down his congregation's woodenness and awaken them to a saving awareness of sin, he does so because this is the best strategy for forcing his Christian readers to acknowledge their own complacent racism. At other times, he inhabits quite different personas. For example, the title page of his autobiography states twice that his identity is determined by his relation to a particular symbolic landscape: not only is the book titled *A Son of the Forest*, it is also subtitled *The Experiences of William Apes, a Native of the Forest*.[109] By using this stock phrase, "a native of the forest," to define his persona, Apess was deploying the image of "the noble savage" which had become "by the nineteenth century a mask through which a Native American could speak" to white audiences about injustice. By associating himself with Nature, he was inhabiting "a locus of alternative value, a source of social protest and identity that serves as a powerful counterweight to the mainstream ideology of progress."[110] Moreover, by calling himself a "'son of the forest,' he identifies himself not only as an Indian, but also as a product of that hallmark of American Methodism, the 'plain-folks camp-meeting.'"[111] In the antebellum era,

Methodism was a working-class denomination to which people of color were welcomed not just as passive worshippers, but as class leaders, exhorters, and even preachers. It was "crucial to [Apess's] self-conception as a Pequot, [since the sect defined] itself as a religion of outsiders pitted against a reprobate elite." Moreover, it allowed him to "articulate to a broad audience a critique of the dominant culture and a vindication of the indigenous one."[112] In short, Apess, like Copway, was a masterful performer of complex identities that shifted fluidly according to the demands of different rhetorical situations.[113]

The "Deep Brown Wilderness" of *A Son of the Forest*

The context for Apess's autobiography was an ecosocial emergency, the seemingly final stage of the two-century process of exiling indigenous people from the land between the Atlantic seaboard and the Mississippi River. That process began in earnest with the 1634–1637 Pequot War in which soldiers from the Massachusetts and Plymouth colonies, along with Narragansett and Mohegan allies, overwhelmed and destroyed the smallpox-weakened Pequot tribe of eastern Connecticut. The Puritan army, led by Captain John Mason, surrounded a Pequot town on the banks of the Mystic (Misistuck) River. The main force of Pequot warriors, led by the sachem Sassacus, was absent, raiding nearby Hartford. Mason ordered his soldiers to set fire to the town and then shoot all of the inhabitants who attempted to escape. At least 600 elders, women, and children were killed. Fewer than 20 survived. After discovering the massacre, the Pequot warriors fled west along the coast of Connecticut. Mason and his army trapped them in a swamp near present-day Fairfield, where they killed or captured all but a few. About 200 scattered members of the tribe surrendered and were sold into slavery. The Massachusetts authorities declared that the name "Pequot" should never be used again. This Mystic Massacre gave the colonists control of the mouth of the Connecticut River, which allowed them access by boat to the fertile valleys of Western Massachusetts. Of course, it was just one more episode in a long story that Apess describes in the appendix to *A Son of the Forest*:

> Ever since the discovery of America by that celebrated navigator, Columbus, the "civilized" or enlightened natives of the old world regarded its inhabitants as an extensive race of "savages!" – of course they were treated as barbarians, and for nearly two centuries they suffered without intermission, as

the Europeans acted on the principal that *might* makes *right* – and if they could succeed in defrauding the natives out of their lands, and drive them from the seaboard, they were satisfied for a time. With this end in view, they sought to "engage them in war, destroy them by thousands with ardent spirits, and fatal disorders unknown to them before." Every European vice that had a tendency to debase and ruin both body and soul was introduced among them. Their avowed object was to obtain possession of the goodly inheritance of the Indian, and in their "enlightened" estimation, the "end justified the means."[114]

In 1829, when Apess wrote this passage, ethnic cleansing had become the law of the land in the United States. President Andrew Jackson had just taken office, after campaigning on his reputation as an "Indian killer," and he immediately began to prosecute an aggressive policy of Indian removal in the South, forcing the tribes of the Five Nations (the Cherokee, Creek, Chickasaw, Choctaw, and Seminole) to relocate to reservations in what is now Oklahoma. Jackson's addresses to Congress describe removal as an inevitable and even natural process:

> Humanity has often wept over the fate of the aborigines of this country, and philanthropy has been long busily employed in devising means to avert it, but its progress has never for a moment been arrested, and one by one have many powerful tribes disappeared from the earth.... Nor is there anything in this which, upon a comprehensive view of the general interests of the human race, is to be regretted. Philanthropy could not wish to see this continent restored to the condition in which it was found by our forefathers. What good man would prefer a country covered with forests and ranged by a few thousand savages to our extensive Republic, studded with cities, towns, and prosperous farms, embellished with all the improvements which art can devise or industry execute, occupied by more than 12,000,000 happy people, and filled with all the blessings of liberty, civilization, and religion?[115]

By the end of Jackson's two terms in office, about 50,000 members of the Five Nations had been driven from their homes in Georgia, Tennessee, and North and South Carolina. At least 4000 died of exposure, starvation, and disease on the forced march to the West that came to be known as the Trail of Tears.

Given this context, we might expect a book entitled *A Son of the Forest* to rely heavily on the native wilderness *topos*. On the contrary, there is almost no landscape description whatsoever in *A Son of the Forest*. If we adjust our expectations, though, we can read Apess's autobiography as environmental literature precisely because it *does not* describe Native Americans living in harmony with Nature. Instead, it displays a photographic negative image. The story takes place in the busy towns, farms, roads, and battlefields of early national New England and Quebec. As the narrative progresses, Apess wanders a racist colonial dystopia, searching for a way to survive. He works for a few weeks at a time as a domestic servant or a manual laborer, often for employers who cheat him of his wages. His time as a soldier in the U.S. Army follows the same pattern, as does his eventual career as a Methodist circuit rider. Throughout the narrative, he is a displaced person who embodies the alienation of indigenous people from their home ground of bodily sustenance and tribal sovereignty. His narrative is haunted by the invisible specter of the land that his Pequot ancestors worked as a common estate before they were uprooted. By refusing to cater to his readers' appetite for inspiring prospects and spiritual dells, and by gesturing instead toward the materiality of the colonized land that once supported his ancestors' bodily life, Apess gives the New England forests a new symbolic meaning. In his sermon "The Increase of the Kingdom of Christ," he coins a phrase, "the deep brown wilderness," which makes an apt term for the radically reframed *topos* that he employs in his autobiography.[116] Instead of William Bradford's hideous and desolate wilderness or James Fenimore Cooper's primitive frontier, *A Son of the Forest* takes place in the white "civilization" that has superimposed itself on George Copway's native wilderness. Rather than addressing Native American removal, as Copway does, by celebrating his tribe's customary lifeways, Apess instead mounts a virtuoso performance of post-removal Pequot identities in a colonized, degraded, and sterile landscape.

The first chapter of *A Son of the Forest* establishes Apess's identity, positions him in the racial hierarchy of early national New England, and diagnoses the present condition of his tribe. He begins by listing the usual autobiographical information: "William Apes, the author of the following narrative, was born in the town of Colereign, Massachusetts, on the thirty-first of January, in the year of our Lord seventeen hundred and ninety-eight." By using the third person, stating the facts precisely, and referring to the Christian calendar, Apess places himself on an equal footing with his reader. He demonstrates that he is oriented to time and place in white society and that his narrative can be checked for accuracy. In the very next sentence, though, he turns to strategic fiction and fancifully traces his ancestry "to the royal family of Philip, king of the Pequod tribe of Indians, so well known in that part of American history, which relates to the wars between the whites and the natives."[117] This passage conflates the two major wars that took place in New England in the seventeenth century: the Pequot War and King Philip's War. Metacom, also known as King Philip, was the leader of the powerful Wampanoag Confederacy that attempted to drive the colonists out of New England in 1675–76. By recasting Metacom as a Pequot leader 40 years earlier and by claiming to be descended from him, Apess characterizes himself as the product of a 200-year history of race war. He describes the colonization of New England as a conflict between two hostile nations that were differentiated by little more than skin color: "the goodly heritage occupied by this once happy, powerful, yet peaceful people was possessed in the process of time by their avowed enemies, the whites, who had been welcomed to their land in that spirit of kindness so peculiar to the red men of the woods."[118] In this passage, Apess establishes a categorical equality between red and white that authorizes native land claims and delegitimizes removal. Apess, the "son of the forest," speaks for the "red men of the woods" to protest the white occupation of the "goodly heritage" of all Native Americans.[119]

However, Apess complicates this narrative of racial conflict by frankly acknowledging the bodily realities of colonization. He notes, "My grandfather was a white man [and my] grandmother was, if I am not misinformed, [King Philip's] granddaughter, and a fair and beautiful woman." This flat statement takes on its full meaning a few sentences later when he observes that the Pequots "were subject to [the] intense and heart corroding affliction ... of having their daughters claimed by the conquerors." The shame and pain of rape were experienced not just by the victims of the wars and their relatives, but by the Pequots as a whole: "[H]owever much subsequent efforts were made to soothe their sorrows, in this particular, they considered the glory of their nation as having departed."[120] With his use of the loaded verb "claimed," Apess suggests that the latter-day Pequots' loss of bodily integrity was deplorable most of all because it eroded the collective identity upon which their land claim depended according to the racialized logic of resource conflict that he has just established. Apess represents *himself* as an embodiment of the Pequots' degradation. He calls himself a "worm of the earth" whose "father was of mixed blood – *his* father being a white man, and his mother, a native of the soil, or in other words a red woman."[121]

After explaining his mixed parentage in this way, Apess bluntly describes the misery and paralysis of the Pequot community into which he was born. And in doing so, he reveals that another kind of pollution has been introduced into Pequot bodies, namely alcohol: "It makes me shudder even at this time, to think how frequent, and how great have been our sufferings in consequence of the introduction of this 'cursed stuff' into our family – and I could wish, in the sincerity of my soul, that it were banished from our land." Apess details how he and his siblings lived on the poorest food and suffered through winter in rags: "Truly, we were in a most deplorable condition – too young to obtain subsistence for ourselves, by the labor of our hands, and our wants almost totally disregarded by those who should have made every exertion to supply them." In response to the children's misery, the "white neighbors took pity on us and measurably administered to our wants by bringing us frozen milk, with which we were glad to satisfy the calls of hunger."[122] This picture of neglect is intensified by a harrowing scene in which the four-year-old Apess's grandmother beats him "most unmercifully with a club" after coming home intoxicated:

> [She] asked me if I hated her, and I very innocently answered in the affirmative as I did not then know what the word meant, and thought all the while that I was answering aright; and so she continued asking me the same question, and I as often answered her in the same way, whereupon she continued beating me, by which means one of my arms was broken in three different places.[123]

The boy's uncle intervenes and decides "to make the whites acquainted with [his] condition." A neighbor, Mr. Furman, comes to investigate, and finds the young boy "dreadfully beaten, and the other children in a state of absolute suffering." Furman appeals to the town selectmen on their behalf, and they are "bound out" to white families as servants. Apess is "entirely disabled in consequence of [his] wounds [and must be] supported at the expense of the town for about twelve months."[124] Forced into dependency on whites by his own grandparents' abuse, Apess is a battered exemplar of his tribe's decline into absolute abjection.[125]

Apess seems at first to accept the idea that his grandmother's behavior can be explained in racial terms. He notes that it was his pure-blooded mother's parents who beat him, while by contrast, his mixed father's parents "were Christians [who] lived and died happy in the love of God." He invites the reader to agree: "I presume that the reader will exclaim, 'what savages your grand parents were to treat unoffending, helpless children in this cruel manner.'" But in response to this rhetorical question, Apess reverses himself and insists that her "unnatural conduct was the effect of some cause." He explains this statement in the following terms:

> I attribute [her violence] in a great measure to the whites, inasmuch as they introduced among my countrymen, that bane of comfort and happiness, ardent spirits – seduced them into a love of it, and when under its unhappy influence, wronged them out of their lawful possessions – that land, where reposed the ashes of their sires; and not only so, but they committed violence of the most revolting kind upon the persons of the female portion of the tribe, who previous to the introduction among them of the arts, and vices, and debaucheries of the whites, were as unoffending and happy as they roamed over their goodly possessions, as any people on whom the sun of heaven ever shown. The consequence was, that they were scattered abroad. Now many of them were seen reeling about intoxicated with liquor, neglecting to provide for themselves and families, who before were assiduously engaged in supplying the necessities of those depending on them for support.[126]

According to Apess, the Pequots have been alienated from the land that was the basis of their collective well-being. As a result, they are incapable of fulfilling their most basic material needs and have forsaken their social responsibilities. Their desolation is the outcome of a two-century race war that began with the introduction of twin toxins, white blood and alcohol, into their bodies.

Despite this harrowing portrait of bodily and cultural disintegration, Apess suggests in this first chapter that the Pequots' historical trajectory can be reversed. He depicts his mixed-race father as a neo-traditionalist who has decisively avowed his native heritage and who has taken steps to reclaim it: "On attaining a sufficient age to act for himself, he joined the Pequod tribe, to which he was maternally connected. He was well received, and in a short time afterwards, married a female of the tribe, in whose veins a single drop of the white man's blood never flowed." Along with this pure-blooded partner, Apess's father "removed to the back settlements" to start a family.[127] Apess himself, in other words, is the product of his parents' deliberate pursuit of bodily, cultural, and geographic healing. In the narrative that follows, he tells the story of his own struggles to overcome the poisons circulating in his body so that he can take up the mantle of his father's program of cultural renewal. In telling this tale, Apess strategically repurposes the homogenizing racial categories of red and white in order to substantiate pan-Indian land claims at precisely the moment when the Jackson administration was pursuing the final displacement of indigenous people east of the Mississippi. Moreover, he represents himself subjunctively as a purified red man who preaches an anti-racist liberation theology that will lead, in the end, to a millennial restoration of native control of the "deep brown wilderness" of America.

THE POLITICS OF METHODISM AND REPUBLICANISM IN *A SON OF THE FOREST*

Since Apess's autobiography is designed to justify the author's membership in the spiritual community that he is addressing, Apess begins his second chapter by establishing the Methodist doctrinal framework that will shape his religious experience: "the Spirit of divine truth ... visits the mind of every intelligent being ... with the design of convincing man of sin and of a judgment to come."[128] Genuine conversion becomes visible "in the Christian deportment of every soul under the mellowing and sanctifying influences of the Spirit of God." Because a person's outward behavior is driven by their inner state, Methodist theology has a significant political dimension. If, as Apess puts it, all people would "yield to the illuminating influences of the Spirit of God – then wretchedness and misery would abound no longer, but every thing of the kind give place, to the pure principles of peace, godliness, brotherly kindness, meekness, charity and love." Unfortunately, the "perverseness of man in this respect, is one of the great and conclusive proofs of ... the rebellious inclination of his unsanctified heart to the will and wisdom of his Creator and his judge."[129] In other words, poverty, hunger, crime, and even genocide are the outward signs of a people's woodenness of spirit and defiance of God's grace. At Methodist camp meetings, where congregants often numbered in the thousands, the work of saving souls was *also* the work of reforming a corrupt society. This democratic and worldly version of Protestant Christianity, with its sharp emphasis on the visibility of grace, sets the stage for a narrative in which what seem to be mundane acts and events take on deep theological and political significance.

Apess's first spiritual experience occurs when he is six years old. Mrs. Furman speaks to him "respecting a future state of existence" and sends him to "the grave yard, where many younger and smaller persons than myself were laid to moulder in the earth."[130] The knowledge that he too might die leads the young Apess to look inward and ask spiritual questions.

Soon the Furmans take him to church for the first time, and he begins what will be a long struggle to achieve settled faith: "After a while I became very fond of attending on the word of God – then again I would meet the enemy of my soul, who would strive to lead me away, and in many instances he was but too successful."[131] As an example, Apess tells a light-hearted story of committing "depredations on a water melon patch belonging to one of the neighbors."[132] But not all temptations are so inconsequential. In this introductory chapter, which sets the direction for the coming narrative of Apess's spiritual growth, racism is the most serious obstacle to virtuous behavior that he faces: "I remember that nothing scarcely grieved me so much [as] to be called by a nick name. If I was spoken to in the spirit of kindness, I would be instantly disarmed of my stubbornness, [but] I thought it disgraceful to be called an Indian; it was considered as a slur upon an oppressed and scattered nation."[133] Throughout *A Son of the Forest*, racism takes its place as one of a suite of worldly injustices (like mockery of his faith and withheld wages) and temptations (like alcohol and property crime) that lead Apess away from a pious life. As he confronts these obstacles, he slowly develops the key traits of temperance and steadiness that bear witness to his successful conversion. Crucially, in order to achieve this transformation, he must purify himself of the bodily pollutants that are responsible for the "wretchedness and misery" of the community in which his life begins.

During his indenture to the Furmans, Apess has spells when he resolves "to mend [his] ways and become a better boy" and others when he allows himself to be led astray.[134] At the age of 11, he conspires with another boy to run away to New London. Their plan is discovered, and Mr. Furman decides to "transfer [his] indentures to Judge Hillhouse for the sum of twenty dollars."[135] Apess immediately runs away from Hillhouse and is transferred again, this time to General William Williams, to whom he is bound for the next four years. During this time, "Methodists began to hold meetings in the neighborhood," and Apess attends them in defiance of Williams and the rest of the "respectable" people in the area, whose "sectarian malice" is ignited by the Methodists' fervor and clamor.[136] Apess's explanation of his motives makes clear that his conversion has everything to do with his experience of racial oppression: "For what cared [the whites] for me? They had possession of the red man's inheritance and had deprived me of liberty; with this they were satisfied and could do as they pleased; therefore, I thought I could do as I pleased, measurably. I therefore went to hear the *noisy Methodists*."[137] Apess's faith finally begins to take hold when one day he feels "convinced that Christ died for all mankind – that age, sect, colour, country, or situation, made no difference." Moreover, he comes to feel sure that he is "included in the plan of redemption with all [his] brethren."[138] After struggling mightily with doubt and confusion, Apess finally experiences a visitation of grace while he is working in the garden: "I lifted up my heart to God, when all at once my burden and fears left me – my heart melted into tenderness – my soul was filled with love – love to God, and love to all mankind.... My love now embraced the whole human family."[139]

While *A Son of the Forest* is primarily a conversion story, it also functions as a protest narrative that denounces racism and calls both for Native American civil rights within the United States and sovereignty on tribal lands. This counter-narrative first emerges in Apess's account of his military service. After running from the Williams family at the age of 15, he makes his way to New York, where he encounters "a sergeant and a file of men who were enlisting soldiers for the United States army."[140] This press-gang plies him with alcohol and regales him with stories about "what a fine thing it was to be a soldier." Despite grave doubts, he enlists. "I could not think," he writes, "why I should risk my life and limbs in fighting for the white man, who had cheated my forefathers out of their land." It is specifically in his limbs, in his body, that he feels the impact of the army's racism. First, he personally experiences the deliberate bodily pollution that drove his grandparents to abuse him: "Too much liquor was dealt out to the soldiers, who got drunk very often.... I have known sober men to enlist, who afterwards became confirmed drunkards, and appear like fools upon the earth. So it was among the soldiers, and what should a child do, who was entangled in their net."[141] In addition to falling under the influence of this insidious culture of alcoholism, Apess is harassed by officers who threaten to "stick [his] skin full of pine splinters, and after having an Indian pow-wow over [him] burn [him] to death."[142] Despite these threats to his racialized body, the young Apess survives and successfully assimilates into his unit. Once they move into action, he emphasizes the soldiers' collective bodily misery during a forced march across open country that has been laid waste by retreating British forces: "The people generally, have no idea of the extreme sufferings of the soldiers on the frontiers during the last war; they were indescribable, the soldiers eat with the utmost greediness raw corn and every thing eatable that fell in their way."[143] Similarly, when the army enters combat, Apess emphasizes its effects on the recruits' bodies:

Their balls whistled around us, and hurried a good many of the soldiers into
the eternal world, while others were most horribly mangled. Indeed they
were so hot upon us, that we had not time to remove the dead as they fell.
The horribly disfigured bodies of the dead – the piercing groans of the
wounded and the dying – the cries for help and succour from those who
could not help themselves – were most appalling. I can never forget it.[144]

Despite the gruesome realism of this portrayal of combat, Apess's tone is
quite proud as he narrates his part in the Battle of Plattsburgh, which he
hails in turn as "a proud day for our country." His experience of combat
authorizes his claim of membership in the American body politic, which in
turn sets the stage for the chapter's final paragraph: "Now, according to
the act of enlistment, I was entitled to forty dollars bounty money, and
one hundred and sixty acres of land. The government also owed me for
fifteen months pay. I have not seen any thing of bounty money, land, or
arrearages, from that day to this."[145] Apess states quite directly that the
Army similarly defrauded "hundreds" of Native Americans who served
during the War of 1812. He condemns the federal government for continuing to deny native veterans "the right of citizenship [and] the privilege
of voting for civil officers."[146] Apess's account of coming of age in the
army serves several rhetorical functions. In the book's conversion narrative, his military service functions as a period of spiritual deadness and
bodily illness that prepares the ground for his later return to a more mature
faith. At the same time, in the book's protest narrative, his story of braving
racist discrimination to become one of the "sons of liberty" grounds his
claim to full citizenship and equality before the law.[147]

Apess strategically reverses stereotypical racial roles throughout *A Son
of the Forest*. For instance, when he describes his childhood anger over
being called an "Indian," he speculates archly about the etymology of the
term: "I have often been led to inquire where the whites received this
word, which they so often threw as an opprobrious epithet at the sons of
the forest. I could not find it in the bible, and therefore come to the conclusion that it was a word imported for the special purpose of degrading
us."[148] At the same time, Apess makes clear that his objection to the term
is not just a matter of pride. Racial epithets and the racist stereotypes that
go with them have warped his own perceptions. As a child who was indentured to white families, he felt a "great fear" of his "brethren of the forest"
that was "occasioned by the many stories [he] had heard of their cruelty
towards the whites – how they were in the habit of killing and scalping

men, women and children."[149] These stories had so "completely ... weaned [him] from the interests and affections" of native people, that he ran in terror one day when shadows in the woods made a group of berry pickers appear to have dark skin. By recalling himself, a red male, fleeing from "a company of white females," he reverses the stock roles in this scene. At first the effect is comic, but the comedy soon modulates into sharp irony when Apess observes that when they told stories of native violence, "the whites did not [reveal] that they were in a great majority of instances the aggressors – that they had imbrued their hands in the life blood of [his] brethren [and] driven them from their once peaceful and happy homes."[150]

Apess often intensifies the effect of reversing racial binaries by doing so within the conceptual frameworks of Protestant Christianity and republican political theory that most of his readers took for granted. For instance, after he musters out of the U.S. Army in Canada, he spends several months as an itinerant laborer, and during this time he is routinely cheated of his wages. While describing this trying time, he stages another role reversal in a brief polemical inset:

> It has been considered as a trifling thing for the whites to make war on the Indians for the purpose of driving them from their country, and taking possession thereof.... But let the thing be changed. Suppose an overwhelming army should march into the United States, for the purpose of subduing it, and enslaving the citizens; how quick would they fly to arms, gather in multitudes around the tree of liberty, and contend for their rights with the last drop of their blood. And should the enemy succeed, would they not eventually rise and endeavour to regain liberty? And who would blame them for it.[151]

By describing colonization in republican terms from an indigenous perspective, Apess forces his white audience to contemplate both the racism and the injustice of Indian removal. Similarly, Apess reverses perspective in an account of an indigenous group living the traditional life on the shores of Lake Ontario. He has contracted to work for a Dutch farmer in the area, and he finds an Anishinaabe band living in the nearby woods. "With this situation I was much pleased," he writes, "my brethren were all around me, and it therefore seemed like home."[152] During his brief stay in this surrogate homeland, Apess is deeply impressed: "I could not but admire the wisdom of God in the order, regularity and beauty of creation. I then turned my eyes to the forest and it appeared alive with its sons and daughters. There appeared to be the utmost order and regularity in their encampment and they held all things in common." In this momentary vision of traditional indigenous life, the forest reflects the civic virtues of the people who inhabit it; both are orderly and beautiful. Apess draws an implicit contrast with the forces that scattered the Pequots and that now threaten this community: "Oh what a pity that this state of things should change. How much better would it be if the whites would act like a civilized people, and instead of giving my brethren of the woods 'rum!' in exchange for their furs, give them food and clothing for themselves and children."[153] The terms of the red/white racial binary have been quietly reversed again, such that the orderly Anishinaabe civilization in the peopled wilderness is threatened by "unholy" invaders. Apess goes on to explain that, from a native point of view, Christian missionaries seem hypocritical when they command people "to love men, deal justly, and walk humbly" while they themselves condone the business of colonization and removal. Apess concludes sarcastically that under these circumstances, "we think the whites need fully as much religious instruction as we do."[154] Throughout *A Son of the Forest*, Apess returns regularly to this strategy of reversal in order to reveal the injustice of Indian removal and the hypocrisy of the whites who enact and support it.

After the early chapters of Apess's autobiography establish that, like the native wilderness, his body has been invaded and despoiled, the later chapters narrate a process of reconquest and reclamation. With the return of spring, the woods "were vocal with the songs of the birds; all nature seemed to smile and rejoice in the freshness and beauty" of the season. The Anishinaabeg in Ontario appear to mirror their environment: "My brethren appeared very cheerful ... and enjoyed themselves in hunting, fishing, [and] basket making."[155] Apess is inspired by this vision of traditional indigenous life to rejoin his Pequot relatives in Massachusetts. The conversion narrative, which has been waiting in the wings of his autobiography during his military service and life in Canada, now returns to center stage. He works his way eastward across New York, encountering racist employers and "wretches" who "blackguard" him "because [he is] an Indian."[156] He persists, struggling to find work and to remain temperate, but listening frequently "to the advice of the wicked [and] drinking too much of the accursed liquor again."[157] When Apess finally reaches Connecticut, "the land of steady habits," his homecoming is joyous, and he vows "to go to work and be steady." Along with temperance and steadiness come self-reliance and courage, so when yet another racist employer tries to beat him with "a cart-stake" instead of paying him a month's wages, Apess makes "him put it down as quick as he had taken it up." His decision to reform himself, in other words, goes hand in hand with a new kind of political resolve: "I had been cheated so often that I determined to have my rights this time, and forever after."[158]

Despite his determination "to persevere in the way of well doing," Apess suffers for months from the conviction that he has "committed the unpardonable sin," but he begins to make progress during a winter that he spends living with his aunt "among my tribe at Groton."[159] Sally George is both enthusiastically pious and self-reliant: "She was the handmaid of the Lord, and being a widow, she rented her lands to the whites, and it brought her in enough to live on."[160] Apess describes his time with her as a season of especially authentic and soulful prayer during three-day camp meetings: "We had no house of divine worship, and believing 'That the groves were God's first temples,' thither we would repair."[161] Soon after feeling "moved to rise and speak" for the first time during a meeting, Apess is baptized in the Methodist church. The turning point in his long and difficult conversion comes during a long night that he spends "shut out from the light of heaven – surrounded by appalling darkness – standing on uncertain ground" after getting lost in a swamp on his way to visit his relatives. In his terror and despair, Apess overcomes his doubts: "I raised my heart in humble prayer and supplication to the father of mercies, and behold he stretched forth his hand and delivered me from this place of danger."[162]

Soon after his deliverance from the swamp, Apess goes to live with his father, and while he is learning to make shoes, he decides to take up the "holy work" of salvation: "While in Colreign the Lord moved upon my heart in a peculiarly powerful manner, and by it I was led to believe that I was called to preach the gospel of our Lord and Saviour Jesus Christ."[163] Apess is drawn to Methodism in particular, not only because of its emphasis on a life of temperance and steadiness but also because of its commitment to the idea that "there is no respect of persons with God."[164] And he pushes the limits of that commitment by insisting that people of color should be welcome not just to worship but to preach. Soon after he receives his call, he begins "exhorting sinners to repentance," even though he knows that some in the audience think he is "nothing but a poor ignorant Indian."[165] Before long, he begins to receive invitations to lead meetings. At one, "a great concourse of people ... come out to hear the indian preach" and "the sons of the devil" throw a hat and sticks at him.

Commenting on this event, he writes, "Now I can truly say that a native of the forest cannot be found in all our country, who would not blush at the bad conduct of many who enjoy in a pre-eminent degree the light of the gospel."[166] Despite the organized opposition of white congregants, he persists in accepting invitations to preach in a widening circle of towns.

With his "native tribe" in Groton, Apess finds a welcome refuge from the racism he encounters during his travels. And in Saybrook, where "a few coloured people ... met regularly for worship," he meets "a woman of nearly the same color" as himself who "bore a pious and exemplary character." They are "united in the sacred bonds of marriage," and they start a family.[167] While this event is glossed in just two sentences, it resolves a central tension that has driven the narrative from the beginning, namely, Apess's attitude toward his own mixed racial inheritance. His matter of fact statement about his wife's skin color reveals that he has decided to carry on his father's project of reversing the pollution of his blood. Rather than explain it directly, Apess allows the significance of this decision to emerge from his proud account of working hard to support his wife and children. The responsibilities of fatherhood give him the strength to remain temperate and the motivation to work steadily both at wage labor and as a circuit exhorter, leading prayer meetings in the homes of the faithful.[168] His conversion narrative culminates with a confident assertion of his success as a preacher throughout eastern Connecticut and Massachusetts: "After spending about two months in Boston, I returned home; then I visited New-Bedford, Martha's Vineyard, and Nantucket, preaching the word wherever a door was opened – and the Lord was not unmindful of me, his presence accompanied me, and I believe that much good was done."[169] Apess brings his autobiography to a close by roundly denouncing "the spirit of prejudice" and by making a bold prediction: "Look brethren, at the natives of the forest – they come, notwithstanding you call them '*savage*,' from the 'east and from the west, the north, and the south,' and will occupy seats in the kingdom of heaven before you."[170] This arresting image – a pan-Indian congregation displacing racist whites and converging before the throne of God – perfectly captures the simultaneously religious and political character of Apess's ministry. Rather than amounting to a surrender to white culture and identity, his conversion to an indigenized Methodism allowed him to purify his body of the twin toxins of white blood and alcohol and to achieve pan-Indian political consciousness.[171]

While the first half of *A Son of the Forest* consists of Apess's autobiographical narrative, the second half of the book tells a history of New England that makes the political dimension of his theological ideas explicit. Apess addresses the central interpretive questions about the era of colonization from a Native American perspective. For instance, in response to the claim that the natives were hostile to the first wave of colonists, Apess writes:

> Who were the first aggressors, and who first imbrued their hands in blood? Not the Indian. No: he treated the stranger as a brother and a friend, until that stranger whom he had received upon his fertile soil, endeavoured to enslave him, and resorted to brutal violence to accomplish his designs. And if they committed excesses, they only followed in the footsteps of the whites, who must blame themselves for provoking their independent and unyielding spirits, and by a long series of cruelty and bloodshed, drove them to arms.[172]

Apess supports each major claim with historical examples, many extracted from the ethnographic treatise, *A Star in the West* (1816), by New Jersey congressman and native rights advocate Elias Boudinot, who was a prominent supporter of the theory that Native Americans had descended from the ten lost tribes of Israel. This idea seems absurd now, but at a time when genocide and conversion defined the spectrum of white opinion on Indian policy, it was a way of countering the claim that Native Americans were subhuman beings who could never successfully assimilate into a civilized society. As Apess writes elsewhere, "If, as many eminent men … believe, the Indians of the American continent are a part of the long lost ten tribes of Israel, have not the great American nation reason to fear the swift judgments of heaven on them for nameless cruelties, extortions, and exterminations inflicted upon the poor natives of the forest?"[173] The appendix to *A Son of the Forest* reprints multiple sections of *A Star in the West*, interspersed with editorial commentary and other materials, including Washington Irving's essay, "Traits of Indian Character," which Boudinot too had reprinted in his own book. The resulting textual collage gives the impression of a lively discussion that Apess moderates as an equal and an expert. The topics of conversation range from the expressiveness of native languages, the fullness of native spiritual traditions, the richness of native generosity, and more. Despite its speculative claims about cognates between the Cherokee and Hebrew languages and about the descent of native religious beliefs from Judaism, and despite the overall focus on converting natives in order to bring them back into the Judeo-Christian fold, the appendix effectively humanizes Native Americans and implicitly authorizes their claims to civil rights, treaty rights, and sovereignty on their ancestral land.[174]

Apess's 1831 sermon, "The Increase of the Kingdom of Christ," articulates his liberation theology in millenarian terms and shows that he saw Methodist Christianity as a tool for building pan-Indian solidarity and power. He begins by condemning social and economic inequality: "The kingdoms of this world, with but few exceptions, are confederacies of wrong; the powerful trespass on the weak; the rich live in luxury and rioting, while the poor are enslaved and doomed to much servile drudgery, without any hope of bettering their condition."[175] Apess radicalizes Methodism's emphasis on the idea that "in the kingdom of Christ, the noble of the earth are on an equality with the poor and humble; no distinctions of birth, except that of being born again, are known." This may sound like ordinary pie-in-the-sky preaching, but he makes explicitly clear that "the kingdom of Christ [exists both] on earth and in heaven." He is not enjoining his audience to wait patiently for a reward in the hereafter. Instead, he reminds them that true faith is revealed by just conduct in the here and now: "Every subject of Christ's kingdom strives to purify himself from all evil. He becomes righteous in his dealings toward his fellow man; and, in accordance with the spirit of the kingdom to which he belongs, he can enslave no man – he can oppress no man." Apess's allusion to slavery makes it clear that the target of his censure is the United States, since a truly Christian republic would honor the principle of human equality, regardless of race. In the just world he invokes, "communities, long disturbed by quarrels and wars, strangely forget to fight and live in gentleness and peace." This means that Native Americans "learn another worship and bow themselves in praise and adoration before the Great Spirit, for the first time revealed to them in the fullness of his glory through a suffering and risen Saviour." It also means that the "white man, who has most cruelly oppressed his red brother ... pours out unavailing tears over the wasted generations of the mighty forest hunters and, now they are almost all dead and buried, begins to pity and lament them."[176] Apess implies that the irreversible "national sin" of genocide can be redeemed only by a new dispensation of racial equality and justice on earth.[177] Finally, Apess uses martial imagery to describe the assembly of an army of "soldiers of the cross" who will "conquer the world for Christ" and make his vision of justice a reality.[178] Crucially, indigenous people will take the lead: "The tribes of the wilderness are in motion. They begin to hear their Saviour's voice, sweetly sounding like the voice of the turtledove among the waving trees of the forest. Arise, ye nations in your strength, and glorify your Redeemer with a voice that all creation may hear – with a song of praise that shall sound like the roar of many waters."[179]

Apess found an opportunity to participate in a local campaign that put his political and theological ideas into practice when his circuit rounds took him to the native village of Mashpee on Cape Cod. His account of the "Mashpee Revolt," in which the town's residents sought self-government and religious liberty, is titled *Indian Nullification of the Unconstitutional Laws of Massachusetts, Relative to the Marshpee Tribe: or, The Pretended Riot Explained* (1835). The book begins with a statement of principle: "That their colour should be a reason to treat one portion of the human race with insult and abuse has always seemed to [me] strange…. God has given to all men an equal right to possess and occupy the earth, and to enjoy the fruits thereof, without any distinction."[180] In support of this general claim, Apess builds a pastiche of texts that reconstructs the discussion in the courts and the press about the Mashpees' history and rights. In this virtual debate, Apess defends his allies and hectors his opponents as the topic ranges from the legality of white title to land in New England, the laws that established state guardianship of the Mashpee, the prevailing spirit of anti-Indian racism in Massachusetts, and native rights to education, elective franchise, religious self-determination, tribal self-government, and armed self-defense. *Indian Nullification* also carries forward Apess's ideas about bodily pollution as primary mechanisms of colonization. The book opens with a sarcastic frontispiece engraving that depicts an overbearing colonist handing a bottle of alcohol to a native man who looks skeptical. The image caption reads, "Manner of Instructing the Indians." Apess comments regularly throughout the book on the poverty and dependency caused by alcoholism, and he represents temperance as a central part of a broad program of pan-Indian sovereignty. Apess also takes time to address the question of racial mixing in a similarly sarcastic tone:

> As soon as we begin to talk about equal rights, the cry of amalgamation is set up, as if men of color could not enjoy their natural rights without any necessity for intermarriage between the sons and daughters of the two races. Strange, strange indeed! … Should the worst come to worst, does the proud white think that a dark skin is less honorable in the sight of God than his own

beautiful hide? … All we ask of them is peace and our rights. We can find wives enough without asking any favors of them. We have some wild flowers among us as fair, as blooming, and quite as pure as any they can show.[181]

As always, Apess's pointed and sharp-witted rhetoric measures this issue of racial discrimination against the plain and simple standards of absolute human equality and universal human rights.[182]

Apess's last published work, "Eulogy on King Philip," is the transcript of an address he delivered twice in January 1836 at Boston's 1300-seat auditorium, The Odeon. Apess praises the memory of his ancestor, Metacom, "the hero of the wilderness" who led a pan-Indian alliance that attempted to drive the Puritan colonists out of New England in 1675.[183] Apess directly compares King Philip's War to the American Revolution that began almost exactly a century later, observing that both were struggles for liberty and independence from foreign tyrants.[184] At one point, Apess delivers a speech within a speech that he attributes to Metacom and that may have been passed down by oral tradition:

> BROTHERS, – You see this vast country before us, which the great Spirit gave to our fathers and us; you see the buffalo and deer that now are our support. – Brothers, you see these little ones, our wives and children, who are looking to us for food and raiment; and you now see the foe before you, that they have grown insolent and bold; that all our ancient customs are disregarded; the treaties made by our fathers and us are broken, and all of us insulted; our council fires put out; our brothers murdered before our eyes, and their spirits cry to us for revenge. Brothers, these people from the unknown world will cut down our groves, spoil our hunting and planting grounds, and drive us and our children from the graves of our fathers, and our women and children will be enslaved.[185]

Metacom's speech, with its powerful imagery of game animals, planting grounds, hungry families, and ancestral graves, shows in the clearest possible terms that indigenous claims to sovereignty are about the right to live on and manage the land in order to sustain both individual bodies and the intergenerational community. Apess's way of contrasting the violence and destruction of the colonial era with the native wilderness *topos* continues to infuse the tradition of indigenous environmental literature that, in North America, extends forward from Apess and Kahgegagahbowh/Copway, through Zitkala-Sa (Lakota), Charles Eastman (Dakota), Black Elk (Oglala), and Sarah Winnemuca (Piaute); to Paula Gunn Allen (Laguna), Louise

Erdrich (Anishinaabe), N. Scott Momaday (Kiowa), Leslie Marmon Silko (Laguna), Gerald Vizenor (Anishinaabe), and James Welch (Blackfeet/ GrosVentre); and on to Sherman Alexie (Spokane/Coeur D'Alene), Jimmy Santiago Baca (Apache), Joy Harjo (Creek), Linda Hogan (Chickasaw), Winona LaDuke (Dakota), and Simon Ortiz (Acoma). If the ecological Indian and the native wilderness are myths, they are the kind of living myths that reliably inspire people to fight for a more just and sustainable world.

NOTES

1. Coronado.
2. Deal.
3. Schell and others.
4. Guha and Alier, 25–27, present an impressive list of additional examples of indigenous environmental campaigns in South America prior to 1998. The contributors to Weaver describe examples of Native American and Canadian First Nations organizing efforts in opposition to nuclear dumping, dam building, oil drilling, and coal mining. LaDuke's powerful book, *All Our Relations*, tells the stories of dozens more indigenous environmental campaigns.
5. Indigenous Environmental Network, "Tar Sands."
6. Idle No More, "The Vision."
7. Idle No More, "The Story."
8. Bennett, n.p.
9. Guha and Alier, xiii–xiv.
10. Guha and Alier, 18. Ali demonstrates that, for indigenous communities in North America, decisions about the environment and development (specifically mining projects) are inseparable from perceptions of sovereignty and self-determination.
11. Guha and Alier, xxi.
12. See Ellingson, 359–372, for an account of the racist rhetoric that environmental organizations used in their attacks on the Makah.
13. Sometimes, the relationship between mainstream and indigenous environmentalism gets more complicated. In northern Arizona, for instance, Save the Confluence, a coalition led by both Anglo and Diné activists, opposed the Diné tribal government's drive to create jobs by authorizing a Phoenix-based, white developer to build a tram, boardwalk, and lodge at the junction of the Colorado and the Little Colorado Rivers in Grand Canyon National Park.
14. Jacoby, 151.
15. Spence describes the removal of native inhabitants from Yosemite, Yellowstone, and Glacier National Parks in the late 19th and early 20th

centuries. Burnham narrates the story of Native Americans and the national parks in the middle and later twentieth century. He shows that "native people are usually considered an embarrassment in the national parks ... because they are poor [and] poverty, much less Indian poverty, has long made an unfortunate contrast with old money—not to mention an unfortunate clash with the contemplation of the sublime that many wilderness visitors seek. For most travelers, Indian communities at the parks have all the photogenic appeal of rural slums" (8–9). Burnham documents the efforts of Native Americans to regain access to ancestral lands on five Western parks from which they have been excluded by the National Park Service. Dowie widens the lens to survey what he calls the "Hundred-Year Conflict between Global Conservation and Indigenous People." Also see Keller and Turek.

16. Cronon argues that adopting this integrative way of thinking will be crucial for an environmental movement that hopes to meet increasingly global challenges: "My own belief is that only by exploring this middle ground will we learn ways of imagining a better world for all of us: humans and nonhumans, rich people and poor, women and men, First Worlders and Third Worlders, white folks and people of color, consumers and producers—a world better for humanity in all of its diversity and for all the rest of nature too. The middle ground is where we actually live. It is where we—all of us, in our different places and ways—make our homes." Porter, 21–30, critiques Cronon's concept of the "middle way" as a liberal fantasy that ignores the reality of unequal power between natives and colonizers and between environmentalists and corporations. While there is some truth to this objection, my belief is that the goal of middle way thinking is to imagine a future in which the binary opposition of wilderness and civilization will have dissolved because it is no longer functional.

17. Weaver, 19.

18. LaDuke, *Reader*, 78.

19. Krech, 16.

20. *The Ecological Indian* elaborates a position stated earlier by conservation biologist Kent Redford in *Cultural Survival Quarterly*. Redford's main point is that indigenous traditional ecological knowledge and methods of subsistence are not necessarily adaptive in a market economy.

21. Ellingson, 342–358, surveys the "ecologically noble savage debate" of the 1990s as part of a broad historical survey of the trope of the Noble Savage. Ellingson argues that the so-called myth of the Noble Savage is a rhetorical trap, since to "either deny or affirm the nobility of savages is to accept the terms of the myth's own construction, and so to affirm the construction of marginalized peoples in terms of their wildness, cruelty,

or inhumanity and assert the superiority of values epitomized in the idealization of class differences into virtues" (374).

22. The contributors to Minnis and Elisens, for instance, argue that the "struggles of Native Americans to adapt to an ecologically diverse continent resulted in a body of knowledge and experiences providing important information about biological diversity, ecological relationships, resource management, and natural products" (15–16).

23. Porter, xix.

24. The phrase "strategic essentialism," coined by Gayatri Spivak, is sometimes used in cultural studies to describe this rhetorical technique. Porter applies this term to the Makah whale hunt, arguing that it was "carried out to articulate cultural autonomy in what is a post- or neo-colonial context" (53).

25. Middleton, 45–64.

26. White House, "Presidential Proclamation."

27. White House, "Statement." The subsequent presidential administration sharply reduced the size of the monument. That decision is being challenged at this time.

28. Taku River Tlingit First Nation and British Columbia, *Wooshtin Yan too.aat.*

29. Evans, n.p.

30. Taku River Tlingit First Nation and British Columbia, *Wóoshtin Wudidaa*, 12.

31. Krech acknowledges that some Native Americans have "actively helped construct [this] image of themselves" since it has proven to be such a powerful tool (20).

32. Taku River Tlingit First Nation and British Columbia, *Wóoshtin Wudidaa*, 7.

33. Taku River Tlingit First Nation and British Columbia, *Wóoshtin Wudidaa*, 13.

34. Weaver, xi.

35. LaDuke, *Reader*, 64.

36. Ruoff, 252.

37. Ruoff, 258.

38. Twenty-five years after Ruoff called attention to Apess and Copway in *Redefining American Literary History*, these two powerful and historically important books remain neglected. By my count, the *MLA International Bibliography* lists only five articles in which *A Son of the Forest* figures as the primary subject and only two that focus primarily on *The Life, History, and Travels of Kah-ge-ga-gah-bowh.* Tim Fulford devotes the brief chapter of *Romantic Indians: Native Americans, British Literature, and Transatlantic Culture 1756-1830* to Copway's writing, arguing that he "asserts his ownership of the land" by means of extensive quotations of British Romantic verse that convey "intense emotional identification with the landscape"

(281). Fulford also notes that Copway "shift[s] his modes of address as he strove to constantly reinvent himself and the Indianness he represented" (283). I differ from Fulford's assessment of the rhetorical power of these strategies, particularly when he claims that Copway achieves no more than a "generalized, feminized, infantilized" native identity because he relies on a "clichéd and sentimentalized derivative" of the Romantic discourse of nature (289–90).

39. My sketch of Copway's life relies heavily on "Kahgegagahboh: Canada's First Literary Celebrity in the United States," Donald Smith's contribution to the 1997 edition of *Life, Letters, and Speeches*. This is the most substantial and exhaustively researched narrative of the life. It builds on an article Smith published earlier in the *Journal of Canadian Studies*. Both versions are useful sources of basic information, but Smith is strangely hostile to his subject, and his interpretive remarks should be evaluated with care.

40. Knobel, 178ff., provides historical context and details on Copway's campaign for a native state in the upper Midwest.

41. Clark, v. *The Ojibway Conquest* appears to have been written by Julius Taylor Clark, an educational agent to the Anishinaabeg. The preface to an 1898 "souvenir edition" of the poem, which was published in Clark's name, explains that the purpose of this arrangement was to "raise funds to aid [Copway] in his work among his people."

42. *Report of the Proceedings*, 42. In "An Ojibwa Conquers Germany" Bernd Peyer gives an ironic though ultimately sympathetic account of Copway's European tour.

43. Flint, 216–220, presents Copway in Europe as a native flaneur who, in *Running Sketches*, records his admiration for Western modernity.

44. Donald Smith, "The Life of George Copway," 5 and 38–39. Smith's reassessment of Copway, 25 years later in *Mississauga Portraits*, is more measured.

45. Walker, 85.

46. Bellin, *Demon*, 191–192.

47. Rex, 2–3.

48. Bellin, *Demon*, 187–199; Hulan, 99–112; Hutchings, *Romantic Ecologies*, 164–175.

49. See https://publications.newberry.org/digitalexhibitions/exhibits/show/indiansofthemidwest/indianlives/item/295

50. See http://blog.nyhistory.org/wp-content/uploads/2016/10/Copway.Portrait.jpg

51. Copway, *Life, History, and Travels*, i.

52. Copway, *Life, History, and Travels*, iii.

53. Copway, *Life, History, and Travels*, vi.

54. Copway, *Life, History, and Travels*, vii.
55. *Exodus* 2:15 and 3:8–10.
56. Copway, *Life, History, and Travels*, 5.
57. Copway, *Life, History, and Travels*, 6.
58. Copway, *Life, History, and Travels*, 8.
59. Copway, *Life, History, and Travels*, 8.
60. Copway, *Life, History, and Travels*, 9.
61. Copway explains: "The Ojebwas are called, here and all around, Massissaugays, because they came from Me-sey Sah-gieng, at the head of Lake Huron, as you go up to Sault St. Marie falls" (*Life, History, and Travels*, 13).
62. Copway, *Life, History, and Travels*, 11.
63. Copway, *Life, History, and Travels*, 20.
64. Copway, *Life, History, and Travels*, 12–13.
65. Copway, *Life, History, and Travels*, 12 and 14.
66. Copway, *Life, History, and Travels*, 12.
67. Copway, *Life, History, and Travels*, 20.
68. Copway, *Life, History, and Travels*, 16.
69. Copway, *Life, History, and Travels*, 16–17.
70. Copway, *Life, History, and Travels*, 17–18.
71. Copway, *Life, History, and Travels*, 21.
72. Copway, *Life, History, and Travels*, 21.
73. Copway, *Life, History, and Travels*, 7.
74. In his contribution to *Transatlantic Literary Ecologies*, Kevin Hutchings briefly discusses Copway's varying rhetorical engagements with the trope of the "Vanishing American" in the context of transatlantic conversations about the idea of extinction (Hutchings and Miller, 64–68).
75. Copway, *Life, History, and Travels*, 23.
76. Copway, *Life, History, and Travels*, 23.
77. Copway, *Life, History, and Travels*, 27.
78. Copway, *Life, History, and Travels*, 34.
79. Copway, *Life, History, and Travels*, 32.
80. Copway, *Life, History, and Travels*, 34–35.
81. Copway, *Life, History, and Travels*, 40.
82. Copway, *Life, History, and Travels*, 41.
83. Copway, *Life, History, and Travels*, 41–42.
84. Copway, *Life, History, and Travels*, 71.
85. Copway, *Life, History, and Travels*, 73.
86. Copway, *Life, History, and Travels*, 76 and 143.
87. Copway, *Life, History, and Travels*, 51 and 82.
88. Copway, *Life, History, and Travels*, 66.
89. Copway, *Life, History, and Travels*, 89.

90. Copway, *Life, History, and Travels*, 188.
91. Copway, *Life, History, and Travels*, 190–191, interpolation in original.
92. Copway, *Life, History, and Travels*, 194.
93. Copway, *Life, History, and Travels*, 199–200.
94. Copway, *Life, History, and Travels*, 198.
95. Copway, *Life, History, and Travels*, 199.
96. Copway, *Life, History, and Travels*, 194 and 202.
97. Copway, *Life, History, and Travels*, 200.
98. Copway, *Life, History, and Travels*, 200–201.
99. Ashwill compares Apess to two other prominent Native American autobiographers who were active at the same time, Elias Boudinot and Black Hawk.
100. McQuaid's appreciative biographical article first brought Apess to the attention of modern scholars and remains a clear-eyed assessment of his historical importance. In "Eulogy on William Apess," Warrior makes a series of fascinating and pointed speculations about the last three years of Apess's life and his decision, as a working writer, to move to New York City, which was quickly displacing Boston as the literary capital of the United States.
101. Apess, *Son*, 60.
102. Apess, *Experiences*, 55.
103. Bizzell reads "An Indian's Looking-Glass for the White Man" and "Eulogy on King Philip" as "mixed blood jeremiads" that use the traditional Puritan polemical sermon to articulate "an Indian viewpoint on New England history, the Christian religion, and the prospects of American democracy" (46).
104. Apess, *Experiences*, 60.
105. Peyer, *The Elder's Wrote*.
106. Peyer's chapter on Apess in *The Tutor'd Mind* remains the best synthetic overview of the available information on Apess's life and provides a succinct introductory overview of his five major published works.
107. Krupat, *Voice*, 145. To his credit, Krupat reverses this assessment in "William Apess: Stories of Survivance," where he argues that, in "An Indian's Looking-Glass for the White Man" and "Eulogy for King Philip," Apess insists "on the ongoing agency and activity of the Native" (103).
108. In "Shadow Casting," Lopenzina surveys Apess's life and writings through the lens of "survivance," which Gerald Vizenor defines as "an active sense of presence, [of] the continuance of native stories.... Native survivance stories are renunciations of dominance, tragedy and victimry" (vii). Sayre rejects the "binary opposition between assimilation and authenticity" that governs judgments of native texts and that exclusively

valorizes orality as opposed to literacy. Similarly, Haynes rejects what she sees as a false choice between singular cultural identities and argues that *A Son of the Forest* displays a cultural "bifocality," such that the multiple dimensions of Apess's subjectivity "are mutually sustaining, interactive, and dialectic" (26). Trodd argues that Apess records a "relational, inter-subjective individuality" and builds "cultural bridges" that negotiate the contact zone (141). Donaldson prefers the term "retraditionalization" to designate the process often referred to as "syncretism," because it "shifts the focus from either the passive imposition or unquestioned acceptance of Euro-American Christianity to much more intricate forms of Native agency" (194 and 196). Barry O'Connell praises Apess as an exemplar of the Native American capacity "to be endlessly inventive, patient, resistant to despair" (99).

109. Until 1836, Apess spelled his surname with a single s. Except where otherwise noted, citations of *A Son of the Forest* refer to the second edition. Gussman offers a historicist account of Apess's revisionary republicanism in "Eulogy for King Philip."

110. Kucich, "Sons of the Forest," 15 and 19.

111. Tiro, 655.

112. Tiro, 654.

113. My reading of Apess aligns closely with Tim Fulford's argument, in *Romantic Indians,* that his "subversive" writing is "shaped by the genres of Protestantism (the sermon as well as the conversion narrative), by the motifs of Romanticism (the rural idyll and the Ossianic sublime) but also by satiric wordplay" (235).

114. Apess, *Son*, 115.

115. Richardson, v. III, 1083–1084.

116. Apess, *On Our Own Ground*, 111.

117. Apess, *Son*, 7.

118. Apess, *Son*, 8.

119. Velikova performs a fascinating rhetorical analysis of Apess's evolving account of his Pequot ancestry and shows that his depictions of New England Indians alternate between abject, though noble, victims (in which case he uses the Pequots as an example) or "sublime and awe-inspiring heroes like King Philip" (318). She argues that Apess's pan-Indianism "strategically privileges racial unity over tribal differences" (326) in order to critique U.S. Indian policy during a time when the public's attention was focused on the Cherokee, and the New England tribes suffered in obscurity. Konkle argues that, by insisting on Indian nationhood, Apess "denies the validity of the concept of inherent Indian difference" (459). In "Red Routes," Bellin responds to Konkle by placing Apess in a prophetic nativist tradition that demanded sovereignty on the basis of claims of fundamental cultural difference.

120. Apess, *Son*, 7–8.
121. Apess, *Son*, 7 and 8–9, emphasis added.
122. Apess, *Son*, 10.
123. Apess, *Son*, 11–12.
124. Apess, *Son*, 12–13.
125. Mielke examines Apess's masterful use of sentimental rhetoric of sympathy in this passage and elsewhere in *A Son of the Forest*.
126. Apess, *Son*, 12.
127. Apess, *Son*, 9.
128. Apess, *Son*, 16. Brooks describes the origins of African American and Native American literature in religious experience in the latter half of the eighteenth century. In a discussion of the writings of Phillis Wheatley, Jupiter Hammon, Samson Occom, John Marrant, and others, she writes, "Their pioneering contributions to American literature came about in connection with [evangelical Christianity], and they reveal how religious formulas such as conversion, revival, and resurrection answered the alienating and mortifying effects of slavery, colonialism, and racial oppression" (9).
129. Apess, *Son*, 16.
130. Apess, *Son*, 18.
131. Apess, *Son*, 20.
132. Apess, *Son*, 24.
133. Apess, *Son*, 20.
134. Apess, *Son*, 26.
135. Apess, *Son*, 31.
136. Apess, *Son*, 37.
137. Apess, *Son*, 38, emphasis in original.
138. Apess, *Son*, 41.
139. Apess, *Son*, 45.
140. Apess, *Son*, 53.
141. Apess, *Son*, 54.
142. Apess, *Son*, 58.
143. Apess, *Son*, 59.
144. Apess, *Son*, 62.
145. Apess, *Son*, 65–66.
146. Apess, *Son*, 66.
147. Bayers analyzes Apess's rhetorical appropriation of classical republican gender norms, including martial valor, in order "to further the rights of Native people" (142).
148. Apess, *Son*, 20–21.
149. Apess, *Son*, 22–23.
150. Apess, *Son*, 23.

151. Apess, *Son*, 67.
152. Apess, *Son*, 69.
153. Apess, *Son*, 70.
154. Apess, *Son*, 71.
155. Apess, *Son*, 73–74.
156. Apess, *Son*, 76–77.
157. Apess, *Son*, 78.
158. Apess, *Son*, 79.
159. Apess, *Son*, 83 and 81.
160. Apess, *Son*, 84.
161. Apess, *Son*, 85. The embedded quotation is from William Cullen Bryant's, "A Forest Hymn," which concludes with a prayer, "Be it ours to meditate,/In these calm shades, thy milder majesty,/And to the beautiful order of thy works/Learn to conform the order of our lives."
162. Apess, *Son*, 90.
163. Apess, *Son*, 92.
164. *Romans* 2:11.
165. Apess, *Son*, 93 and 92.
166. Apess, *Son*, 94.
167. Apess, *Son*, 97–98.
168. In addition to confronting racism in the pews, Apess also faces stiff opposition from the clergy. The first edition of *A Son of the Forest* includes a section that Apess struck from the second edition, in which the "monarchial" Methodist Episcopal Church refuses to license him to preach, so he leaves in protest, and joins the "republican" Methodist Society (115).
169. Apess, *Son*, 104.
170. Apess, *Son*, 116.
171. Miller gives a richly detailed and nuanced historical account of Apess's complex relationship to Methodism and his appropriation of Methodist temperance discourse.
172. Apess, *Son*, 119.
173. Apess, *On Our Own Ground*, 106.
174. Gustafson examines how Apess uses the lost tribes thesis and other elements of antebellum prophetic discourse to authorize claims to Native American cultural autonomy.
175. Apess, *On Our Own Ground*, 102.
176. Apess, *On Our Own Ground*, 102.
177. Apess, *On Our Own Ground*, 107.
178. Apess, *On Our Own Ground*, 108.
179. Apess, *On Our Own Ground*, 109.
180. Apess, *Indian Nullification*, 10.
181. Apess, *Indian Nullification*, 103–104.

182. In "William Apess's Nullifications," Kucich argues that *Indian Nullification* "is a hybrid text, firmly rooted in Mashpee culture and fully engaged with Jacksonian America" and that it "is an exemplary text of the contact zone, the space in which people from different cultures, locked in the upheavals and exploitations of colonial conflict, can negotiate the terms of their difference" (12–13). Gaul suggests that the book "enacts the very model of toleration and open dialogue that [Apess] seeks in the country's public life" because it allows "debate to emerge within [the] text by placing pieces of varying viewpoints side by side" (275–276).

183. Apess, *Eulogy*, 6.

184. Stevens observes that in "Eulogy on King Philip," Apess creates not only "a revisionist appraisal of King Philip, but an alternative [American] historiography from the perspective of a Native American" (79–80). In "What to the American Indian is the Fourth of July?," Lopenzina calls Apess's speech "an act of revisionist historiography" (675). Wolfe describes the speech as an act of "rhetorical sovereignty" aimed at refuting the "the myth of Indians as the 'Vanishing Americans'" in part by redirecting the language of mourning and melancholia in his narrative of King Philip's War (2).

185. Apess, *Eulogy*, 35–36.

Bibliography

Ali, Saleem H. 2003. *Mining, the Environment, and Indigenous Development Conflicts*. Tucson: University of Arizona Press.

Apess, William. 1829. *A Son of the Forest: The Experiences of William Apes, a Native of the Forest, Comprising a Notice of the Pequod Tribe of Indians. Written by Himself*. New York: Published by the Author. *Internet Archive*.

———. 1831. *A Son of the Forest: The Experiences of William Apes, a Native of the Forest, Comprising a Notice of the Pequod Tribe of Indians. Written by Himself*. 2nd ed. New York: Published by the Author. *Internet Archive*.

———. 1833. *The Experiences of Five Christian Indians of the Pequot Tribe*. Boston: James B. Dow. *Internet Archive*.

———. 1835. *Indian Nullification of the Unconstitutional Laws of Massachusetts, Relative to the Marshpee Tribe: or, The Pretended Riot Explained*. Boston: J. Howe. *HathiTrust*.

———. 1836. *Eulogy on King Philip*. Boston: By the Author. *Google Books*.

———. 1991. In *On Our Own Ground: The Complete Writings of William Apess, a Pequot*, ed. Barry O'Connell. Amherst: University of Massachusetts Press.

Ashwill, Gary. 1994. Savagism and Its Discontents: James Fenimore Cooper and His Native American Contemporaries. *American Transcendental Quarterly* 8 (3): 211–227.

Bayers, Peter L. 2006. William Apess's Manhood and Native Resistance in Jacksonian America. *Melus* 31 (1): 123–146.

Bellin, Joshua David. 2001. *The Demon of the Continent: Indians and the Shaping of American Literature*. Philadelphia: University of Pennsylvania Press.

———. 2010. Red Routes: William Apess and Nativist Prophecy. *Literature in the Early American Republic: Annual Studies on Cooper and His Contemporaries* 2: 45–80.

Bennett, Caroline. Patricia Gualinga: Warrior for the Amazon. *Eye on the Amazon*. Amazon Watch. http://amazonwatch.org/news/2013/1120-patricia-gualinga-warrior-for-the-amazon

Bizzell, Patricia. 2006. (Native) American Jeremiad: The 'Mixedblood' Rhetoric of William Apess. In *American Indian Rhetorics of Survivance: Word Medicine, Word Magic*, ed. Ernest Stromberg, 34–49. Pittsburgh: University of Pittsburgh Press.

Brooks, Lisa. 2008. *The Common Pot: The Recovery of Native Space in the Northeast*. Minneapolis: University of Minnesota Press.

Burnham, Philip. 2000. *Indian Country, God's Country: Native Americans and the National Parks*. Washington, DC: Island.

Clark, Julius Taylor. 1898. *The Ojibue Conquest, an Indian Episode, with Other Waifs of Leisure Hours*. Topeka: Kansas. *Internet Archive*.

Copway, George. 1847. *The Life, History, and Travels of Kah-ge-ga-gah-bowh (George Copway), a Young Indian Chief of the Ojibwa Nation, a Convert to the Christian Faith, and a Missionary to His People for Twelve Years*. Albany: Weed and Parsons. HathiTrust.

———. 1850. *Organization of a New Indian Territory, East of the Missouri River*. New York: S. W. Benedict. *HathiTrust*.

———. 1851a. *Running Sketches of Men and Places, in England, France, Germany, Belgium, and Scotland*. New York: J. C. Riker. *Internet Archive*.

———. 1851b. *The Traditional History and Characteristic Sketches of the Ojibway Nation*. Boston: Benjamin B. Mussey. Internet Archive.

———. 1997. In *Life, Letters, and Speeches*, ed. A. LaVonne Brown Ruoff and Donald B. Smith. Lincoln: University of Nebraska Press.

Coronado, Chris. 2014. 'Last American Indian' Finds Challenges in Performance Art. *Washington Post*. 14 February. https://www.washingtonpost.com/lifestyle/magazine/last-american-indian-finds-challenges-in-performance-art/2014/02/13/08b88100-82ba-11e3-8099-9181471f7aaf_story.html

Cronon, William. 1996. The Trouble with Wilderness. *William Cronon*. http://www.williamcronon.net/writing/Trouble_with_Wilderness_Main.html

Deal, Gregg. Last American Indian on Earth. Gregg Deal. http://greggdeal.com/The-Last-American-Indian-On-Earth.

Donaldson, Laura E. 2005. Making a Joyful Noise: William Apess and the Search for Postcolonial Method(ism). *Interventions* 7 (2): 180–198.

Dowie, Mark. 2011. *Conservation Refugees: The Hundred-Year Conflict Between Global Conservation and Native Peoples*. Boston: MIT Press.

Ellingson, Terry Jay. 2001. *The Myth of the Noble Savage*. Berkeley: University of California Press.

Evans, Bryan. 2014. Personal Correspondence, 30 September.

Flint, Kate. 2009. *The Transatlantic Indian, 1776–1930*. Princeton: Princeton University Press.

Fulford, Tim. 2006. *Romantic Indians: Native Americans, British Literature, and Transatlantic Culture 1756–1830*. New York: Oxford University Press.

Gaul, Theresa Strouth. 2001. Dialogue and Public Discourse in William Apess's *Indian Nullification*. *American Transcendental Quarterly* 15 (4): 275–292. EBSCOHost.

Guha, Ramachandra, and Juan Martinez Alier, eds. 1998. *Varieties of Environmentalism: Essays North and South*. London: Earthscan.

Gussman, Deborah. 2004. O Savage, Where Are Thou?': Rhetorics of Reform in William Apess's 'Eulogy on King Philip. *New England Quarterly* 77 (3): 451–477.

Gustafson, Sandra. 1994. Nations of Israelites: Prophecy and Cultural Autonomy in the Writings of William Apess. *Religion and Literature* 26 (1): 31–53. JSTOR.

Haynes, Carolyn. 1996. 'A Mark for them All to … Hiss At': The Formation of Methodist and Pequot Identity in the Conversion Narrative of William Apess. *Early American Literature* 31: 25–44.

Hulan, Shelly. 2010. Telling a Better Story: History Fiction, and Rhetoric in George Copway's *Traditional History and Characteristic Sketches of the Ojibway Nation*. In *National Plots: Historical Fiction and Changing Ideas of Canada*, ed. Andrea Cabajsky and Brett Josef Grubisic, 99–112. Waterloo: Wilfrid Laurier University Press.

Hutchings, Kevin. 2009. *Romantic Ecologies and Colonial Cultures in the British-Atlantic World, 1770–1850*. Montreal: McGill-Queen's University Press.

Hutchings, Kevin, and John Miller, eds. 2017. *Transatlantic Literary Ecologies: Nature and Culture in the Nineteenth-Century Anglophone Atlantic World*. London: Routledge.

Idle No More. The Story. http://www.idlenomore.ca/story

———. The Vision. http://www.idlenomore.ca/vision

Indigenous Environmental Network. Tar Sands. http://www.ienearth.org/what-we-do/tar-sands/

Jacoby, Karl. 2014. *Crimes Against Nature: Squatters, Poachers, Thieves, and the Hidden History of American Conservation*. Berkeley: University of California Press.

Keller, Robert H., and Michael F. Turek. 1999. *American Indians and National Parks*. Tucson: University of Arizona Press.

Knobel, Dale T. 1984. Know-Nothings and Indians: Strange Bedfellows? *Western Historical Quarterly.* 15, 175–12, 198. *JSTOR.*

Konkle, Maureen. 1997. Indian Literacy, U. S. Colonialism, and Literary Criticism. *American Literature: A Journal of Literary History, Criticism, and Bibliography* 69 (3): 457–486.

Krech, Shephard. 1999. *The Ecological Indian: Myth and History.* New York: Norton.

Krupat, Arnold. 1989. *The Voice in the Margin: Native American Literature and the Canon.* Berkeley: University of California Press.

———. 2008. William Apess: Storier of Survivance. In *Survivance: Narratives of Native Presence,* ed. Gerald Vizenor, 103–121. Lincoln: University of Nebraska Press. Google Books.

Kucich, John. 2006. Sons of the Forest: Environment and Transculturation in Jonathan Edwards, Samson Occom and William Apess. In *Assimilation and Subversion in Earlier American Literature,* ed. Robin DeRosa, 5–23. Newcastle upon Tyne: Cambridge Scholars.

Kucich, John J. 2009. William Apess's Nullifications: Sovereignty, Identity and the Mashpee Revolt. In *Sovereignty, Separatism, and Survivance: Ideological Encounters in the Literature of Native North America,* ed. Benjamin D. Carson, 1–16. Newcastle upon Tyne: Cambridge Scholars.

LaDuke, Winona. 1999. *All Our Relations: Native Struggles for Land and Life.* Cambridge: South End Press.

———. 2002. *The Winona LaDuke Reader: A Collection of Essential Writings.* Stillwater: Voyageur.

Lopenzina, Drew. 2010a. Shadow Casting: William Apess, Survivance, and the Problem of Historical Recovery. In *Gerald Vizenor: Texts and Contexts,* ed. Deborah L. Madsen and A. Robert Lee, 208–230. Albuquerque: University of New Mexico Press. EBSCOHost.

———. 2010b. What to the American Indian Is the Fourth of July? Moving Beyond Abolitionist Rhetoric in William Apess's Eulogy on King Philip. *American Literature: A Journal of Literary History, Criticism, and Bibliography* 82 (4): 673–699.

McQuaid, Kim. 1977. William Apes, Pequot: An Indian Reformer in the Jackson Era. *New England Quarterly* 50 (4): 605–625. JSTOR.

Middleton, Beth Rose. 2011. *Trust in the Land: New Directions in Tribal Conservation.* Tucson: University of Arizona Press.

Mielke, Laura L. 2002. 'Native to the Question': William Apess, Black Hawk, and the Sentimental Context of Early Native American Autobiography. *American Indian Quarterly* 26 (2): 246–270.

Miller, Mark J. 2010. 'Mouth for God': Temperate Labor, Race, and Methodist Reform in William Apess's A Son of the Forest. *Journal of the Early Republic* 30 (2): 225–251.

Minnis, Paul E., and Wayne J. Elisens, eds. 2000. *Biodiversity and Native America.* Norman: University of Oklahoma Press.

O'Connell, Barry. 1993. William Apess and the Survival of the Pequot People. In *Algonkians of New England, Past and Present,* ed. Jane Montague Benes and Peter Benes, 89–100. Boston: Boston University Press.

Peyer, Bernd. 1982. *The Elders Wrote: An Anthology of Early Prose by North American Indians, 1768–1931.* Berlin: Reimer.

———. 1997. *The Tutor'd Mind: Indian Missionary-Writers in Antebellum America.* Amherst: University of Massachusetts Press.

———. 2002. An Ojibwa Conquers Germany. In *Germans and Indians: Fantasies, Encounters, Projections,* ed. Colin G. Galloway, Gerd Gemünden, and Susanne Zantop, 141–164. Lincoln: University of Nebraska Press.

Porter, Joy. 2012. *Land and Spirit in Native America.* Santa Barbara: Praeger. Republished as *Native American Environmentalism: Land, Spirit, and the Idea of Wilderness.* Lincoln: University of Nebraska Press, 2014.

Redford, Kent. 1991. The Ecologically Noble Savage. *Cultural Survival Quarterly* 15 (1). https://www.culturalsurvival.org/publications/cultural-survival-quarterly/ecologically-noble-savage

Report of the Proceedings of the Third General Peace Congress. 1851. London: Charles Gilpin. *Google Books.*

Rex, Cathy. 2006. Survivance and Fluidity: George Copway's *The Life, History, and Travels of Kah-ge-ga-gah-bowh. Studies in American Indian Literatures* 18 (2): 1–33.

Richardson, James, ed. 1897. *Messages and Papers of the Presidents.* New York: Bureau of National Literature. HathiTrust.

Ruoff, A. LaVonne Brown. 1990. Three Nineteenth-Century American Indian Autobiographers. In *Redefining American Literary History,* ed. A. LaVonne Brown Ruoff and Jerry W. Ward Jr., 251–269. New York: Modern Language Association of America.

Sayre, Gordon. 1996. Defying Assimilation, Confounding Authenticity: The Case of William Apess. *Auto/Biography Studies* 11 (1): 1–18.

Schell, Lawrence M., et al. 2013. Relationships of Polychlorinated Biphenyls and Dichlorodiphenyl-dichloroethylene (p,p'-DDE) with Testosterone Levels in Adolescent Males. *Environmental Health Perspectives* 122 (3): 304–309.

Smith, Donald B. 1988. The Life of George Copway or Kah-ge-ga-gah-bowh (1818–1869) – And a review of his writings. *Journal of Canadian Studies* 23 (3): 5–38.

———. 2013. *Mississauga Portraits: Ojibwe Voices from Nineteenth-Century Canada.* Toronto: University of Toronto Press.

Spence, Mark David. 1999. *Dispossessing the Wilderness: Indian Removal and the Making of the National Parks.* New York: Oxford University Press.

Stevens, Scott Manning. 1997. William Apess's Historical Self. *Northwest Review* 35 (3): 67–84.

Taku River Tlingit First Nation and Province of British Columbia. Wóoshtin Wudidaa Atlin Taku Land Use Plan. 19 July 2011a.

———. Wóoshtin Yan too.aat Land and Resource Management and Shared Decision Making Agreement. 19 July 2011b.

Tiro, Karim M. 1996. Denominated 'SAVAGE': Methodism, Writing, and Identity in the Works of William Apess, a Pequot. *American Quarterly* 48 (4): 653–679.

Trodd, Zoe. 2007. Hybrid Constructions: Native Autobiography and the Open Curves of Cultural Hybridity. In *Reconstructing Hybridity: Post-Colonial Studies in Transition*, ed. Joel Kuortti and Jopi Nyman, 139–162. Amsterdam: Rodopi.

Velikova, Roumiana. 2002. 'Philip, King of the Pequots': The History of an Error. *Early American Literature* 37 (2): 311–335.

Vizenor, Gerald. 1999. *Manifest Manners: Narratives on Postindian Survivance*. Lincoln: University of Nebraska Press.

Walker, Cheryl. 1997. *Indian Nation: Native American Literature and Nineteenth-Century Nationalisms*. Durham: Duke University Press.

Warrior, Robert. 2004. Eulogy on William Apess: Speculations on His New York Death. *Studies in American Indian Literatures* 16 (2): 1–13. JSTOR.

Weaver, Jace, ed. 1996. *Defending Mother Earth: Native American Perspectives on Environmental Justice*. Maryknoll/New York: Orbis.

White House. 2016a. Presidential Proclamation – Establishment of the Bears Ears National Monument, 28 December. https://obamawhitehouse.archives.gov/the-press-office/2016/12/28/proclamation-establishment-bears-ears-national-monument

———. 2016b. Statement by the President on the Designation of Bears Ears National Monument and Gold Butte National Monument, 28 December. https://obamawhitehouse.archives.gov/the-press-office/2016/12/28/statement-president-designation-bears-ears-national-monument-and-gold

Wolfe, Eric A. 2008. Mourning, Melancholia, and Rhetorical Sovereignty in William Apess's Eulogy on King Philip. *Studies in American Indian Literatures: The Journal of the Association for the Study of American Indian Literatures* 20 (4): 1–23. Project Muse.

CHAPTER 4

The Green City

INTERSECTIONAL FEMINISM AND THE PUBLIC ENVIRONMENT

In recent years, conversations on social media have condemned mansplaining and ridiculed manspreading. The hashtag #MeToo has demonstrated that sexual harassment and violence have impacted women in everyone's network of friends, colleagues, and acquaintances. Coordinated efforts to expose sexual predators have removed dozens of formerly invulnerable men from positions of power. When a misogynist presidential candidate bragged about grabbing women by their genitals, activists across the United States donned pink pussy hats and T-shirts blazoned with slogans like "Pussy Grabs Back!" On January 21, 2017, half a million people participated in the Women's March on Washington, which was one of the largest demonstrations in the national capital since the Vietnam era and was supported by solidarity protests in hundreds of other cities around the world. After a long backlash, feminism is resurgent.

The "Guiding Vision and Definition of Principles" that was published by the organizers of the Women's March states a remarkably wide range of political commitments: reproductive freedom, passage of an Equal Rights Amendment, LGBTQIA rights, worker's rights and economic justice, immigrant and refugee rights, and environmental health and safety, including "clean water, clean air, and access to and enjoyment of public lands." The manifesto is remarkably direct about the integration of political issues that have often been seen as disconnected or even at odds with one

© The Author(s) 2019

L. Newman, *The Literary Heritage of the Environmental Justice Movement*, Literatures, Cultures, and the Environment, https://doi.org/10.1007/978-3-030-14572-9_4

another: "We must create a society in which women, in particular women – in particular Black women, Native women, poor women, immigrant women, Muslim women, and queer and trans women – are free and able to care for and nurture their families, however they are formed, in safe and healthy environments free from structural impediments."[1] This integrative political philosophy, which is informed by the concept of intersectionality, visibly inspired the millions of marchers in Washington, DC. While some of their hand-lettered signs bore inspirational slogans, such as "Fight Like a Girl" and "Pussy Power," and others carried statements of individual purpose and identification, like "I march for my daughter" and "Men for women's rights," many of the signs made statements about connectivity. Some connected a single political concern to feminism, such as "Gun Violence Is a Women's Issue" or "A Good Planet Is Hard to Find." Others made much more inclusive arguments, such as "Gender Justice = Racial Justice = Economic Justice" and "Women's Rights Are Human Rights: Black, Worker, Immigrant, Trans, Poor, Muslim, Queer, Native, Disabled." An intersectional list of slogans that first appeared at the Women's March has since become common on yard signs, T-shirts, and social media banners as a statement of generalized progressive commitment: "Black Lives Matter," "Women's Rights Are Human Rights," "Climate Change Is Real," "Build Bridges Not Walls," "Love Is Love," "Kindness Is Everything." Perhaps the most comprehensive and pithy slogan of intersectionality that was displayed on the March is "Respect Existence or Expect Resistance."[2]

One way to interpret the mobilizing power of the idea of intersectionality is to suggest that previously disconnected activist communities see the need to converge into a broad alliance against a common adversary. Thus, while any individual demonstration has primary organizers who define a central issue or goal, all of the thousands of marches and rallies that happen every year have merged into a single continuous progressive movement. Another way to inflect this interpretation is to say that a wide range of affinity groups have converged around a resurgent feminist core to form a broad movement that is *environmentalist* in a radically expanded sense of the word. This intersectional alliance seeks to change our *total* ecosocial environment—physical and discursive, legal and cultural, gendered and raced, natural and built.[3] The movement's marches, rallies, and demonstrations may be seen as reinvented or repurposed political tactics, new versions of the general strike for an era when people affiliate and organize according to their identities and positionalities, not their trades and occupations. Marchers bodily reoccupy public spaces from which they and/or their allies have been excluded by the threat of sexual and racial violence.

As philosopher Judith Butler writes, when diverse groups of people take up visible positions in public space, they demonstrate to the world that "it is *this* body, and *these* bodies, that require employment, shelter, health care, and food...; it is *this* body, or *these* bodies, or bodies *like* this body or these bodies, that live the condition of an imperiled livelihood, decimated infrastructure, accelerating precarity."[4] More than protest the ways our shared life world has been degraded, intersectional public assemblies also build and rebuild a possibilistic world on the streets and in the squares that we traverse every day. As people of all kinds gather to voice their varied commitments, there is a phase shift in reality, and for a few hours, an anticipatory landscape shimmers through the surfaces of familiar urban space. There is a carnivalesque suspension of dominant social norms, prohibitions, and rules. Autonomous zones like the Occupy Wall Street encampment take shape within contemporary neoliberal geography, and in these unstable subjunctive spaces, there is a measure of safety, a temporary reprieve from the oppressive violence of hate and the state. Marginalized people are able to live their identities, occupy their chosen positions, and make radical claims with a new degree of freedom, though all recognize that this freedom may be rescinded by police violence at any time. In other words, the intersectional, integrative demonstrations of the present moment perform a powerful *topos*, a landscape of revolution with a long history: the green city.

The green city is a possibilist vision of a gender-inclusive, multicultural, nonviolent, just, and healthy urban community that stands in direct contrast to the sexist, white supremacist, unequal, cruel, and harmful cities that define modernity. If the radical pastoral, the revolutionary sublime, and the native wilderness were landscapes of revolution created in response to the emergence of capitalism and colonialism in a predominantly rural and wild world, the vision of the green city has arisen within and against the new hubs of labor, manufacturing, commerce, transportation, crime, pollution, and disease that have developed around the planet. As of 2008, more than half of the world's people live in urban spaces. In these metropolises, human bodies are traversed by flows of energy, matter, information, and ideas, and it becomes clear that physical, spiritual, and mental wellness are entangled. Physical and temporal proximity make the connections between political issues immediately visible. New configurations of community and even solidarity form fluidly and sometimes dissolve as people live side by side. In the urbanized world that more and more of us now inhabit, the green city is a *topos* that motivates and shapes emergent forms of ecosocial activism.

MARY WOLLSTONECRAFT, EDUCATION, AND THE BODY

The *topos* of the green city emerged within the transatlantic feminist tradition that emerged in the sixteenth and seventeenth centuries with the publication of texts like Jane Anger's *Her Protection for Women* (1589), Anne Bradstreet's *The Tenth Muse Lately Sprung Up in America* (1650), and Mary Astell's *A Serious Proposal to the Ladies for the Advancement of Their True and Greatest Interest* (1694). These texts prepared the way for the great manifesto of the revolutionary era, Mary Wollstonecraft's *A Vindication of the Rights of Women* (1792), which had much greater impact than its predecessors, in part because its rights-based arguments were so well suited to the ideological environment of the period and in part because William Godwin's 1798 memoir of Wollstonecraft made her internationally infamous. As she relentlessly condemned sexual inequality, Wollstonecraft laid the groundwork for the green city by suggesting that the right to access public space was fundamental, since women's bodily and mental well-being are deeply intertwined.

A Vindication starts from the idealist position that sex is insignificant from a divine perspective, because it is a feature of our merely physical bodies, while our immaterial souls are gender-neutral. The godly gift that makes us human is our ability to think: "In what does man's pre-eminence over the brute creation consist? The answer is as clear as that a half is less than the whole; in Reason."[5] Since it is souls that think, not bodies, there can be no difference between men and women in their potential for rationality, and "there is but one way appointed by Providence to lead *mankind* to either virtue or happiness."[6] In *A Vindication*, Wollstonecraft consistently focuses on the disparity between, on the one hand, the equality of disembodied souls that was established by divine providence and, on the other, the oppressive sexual hierarchies that have been established by men. "God," she writes, "has made all things right; but man has sought him out many inventions to mar the work."[7] More specifically, in order to establish male dominance, men have interposed themselves between God and women, usurping His preeminence. Wollstonecraft remarks wryly that if "women were destined by Providence to acquire human virtues ... they must be permitted to turn to the fountain of light, and not [be] forced to shape their course by the twinkling of a mere satellite."[8] Like so many other women writers and writers of color in her time, she works within the universal rights discourse of the Enlightenment, and she argues that social inequalities that are based on bodily differences are unjust, since what makes us all human is immaterial.

Wollstonecraft traces the ongoing oppression of women to the way that they are taught in childhood: "the neglected education of my [female] fellow-creatures is the grand source of the misery I deplore."[9] False beliefs about the differing capacities of boys and girls have led to "a false system of education" in which boys are raised to become independent men in the world, while parents are "more anxious to make [girls into] alluring mistresses than rational wives."[10] In response to this inequity, Wollstonecraft argues that women deserve an education that develops their intellectual and moral self-reliance.[11] The "most perfect education … is such an exercise of the understanding as is best calculated … to enable the individual to attain such habits of virtue as will render it independent."[12] Rather than strive merely to please men, women should seek "to unfold their own faculties" and to "attain conscious dignity by feeling themselves only dependent on God."[13]

Strangely, Wollstonecraft states that she does not "wish to invert the order of things" and that she is only interested in intellectual and moral equality, since she believes that physical equality is out of the question: "I will allow that bodily strength seems to give man a natural superiority over woman; and this is the only solid basis on which the superiority of the sex can be built. But I will insist, that not only the virtue, but the *knowledge* of the two sexes should be the same in nature, if not in degree."[14] Here she seems to accept a sharp delimitation of women's freedom by ceding bodily independence in the material world to men, but other passages in *A Vindication* show that she viewed the body and mind as integrated and that she believed women should "be free in a physical, moral, and civil sense."[15] This comprehensive way of thinking about women's autonomy surfaces repeatedly: "I wish to persuade women to endeavour to acquire strength, both of mind and body."[16] In fact, Wollstonecraft argues that physical and mental well-being are directly linked: "I find that strength of mind has, in most cases, been accompanied by superior strength of body."[17] This linkage has important consequences, since one of most powerful ways that men maintain a rigid gender hierarchy is to constrain women's movement: "To preserve personal beauty, woman's glory, the limbs and faculties are cramped … and the sedentary life which they are condemned to live, while boys frolic in the open air, weakens the muscles and relaxes the nerves."[18] Wollstonecraft implies that, because women have been confined to domestic spaces, they are unable to engage in healthy developmental play. Because they cannot experience the "bodily inconveniences and exertions that are requisite to strengthen the mind," their mental growth is inhibited.[19] She even imagines the mind as stunted by confinement within a body that is in turn confined by gender: "Taught from their infancy that beauty is woman's sceptre, the mind shapes itself to the body, and, roaming round its gilt cage, only seeks to adorn its prison."[20]

Wollstonecraft calls for education to become "a grand national concern" and for the state to fund coeducational day schools that would serve all children, rich and poor.[21] Boys and girls would study side by side, learning about "botany, mechanics, and astronomy," as well as "reading, writing, arithmetic, natural history, and some simple experiments in natural philosophy." The curriculum would also include "the elements of religion, history, the history of man, and politics." All subjects would "be taught, by conversations, in the socratic form." However, "these pursuits should never encroach on gymnastic plays in the open air."[22] This vision of a national system of public education concludes a consistent thread of commentary in *A Vindication* on the connection between bodily health and mental well-being. For instance, as part of her refutation of Jean Jacques Rousseau's ideas of gender-based child-rearing, Wollstonecraft argues that the "baneful consequences which flow from inattention to [girls'] health during infancy, and youth, extend further than is supposed – dependence of body naturally produces dependence of mind."[23] She insists that girls should be "allowed to take sufficient exercise and not confined in close rooms till their muscles are relaxed, and their powers of digestion destroyed."[24] Finally, she argues that the freedom to move bodily through the world will allow women to attain independence of mind: "Let their faculties have room to unfold, and their virtues to gain strength, and then determine where the whole sex must stand in the intellectual scale."[25] If women were granted freedom of access to ideas, information, *and* the world around them, they would develop to their full physical and mental potential, and deep social change would soon follow: "It is time to effect a revolution in female manners – time to restore to them their lost dignity – and make them, as a part of the human species, labour by reforming themselves to reform the world."[26]

A Vindication was republished in Boston and Philadelphia within months after the first London edition appeared in 1792. American readers, who had been prepared by Judith Sargent Murray's famous essay, "On the Equality of the Sexes" (1790), hungrily consumed Wollstonecraft's writings, along with the scandalous details of her romantic life that emerged after her death.[27] In the subsequent decades, ideas about women's bodily freedom circulated widely. As the common school movement gathered momentum in the 1830s, Catharine Beecher advocated widely for daily physical education for girls as well as for boys. Popular conduct books reinforced the importance of outdoor activities for girls. For instance, in *The Mother's Book* (1831), Lydia Maria Child writes: "Amusements and employments which lead to exercise in the open air have greatly the advantage of all others" since they produce "health and cheerful spirits," and she insists that "[g]ardening, sliding, skating, and snow-balling, are all as good for girls as for boys."[28] Child replanted these ideas in her historical and journalistic writings of the 1840s, where they took root and flowered into the *topos* of the green city.

ENVIRONMENTAL FEMINISM IN *THE HISTORY OF THE CONDITION OF WOMEN*

Lydia Maria Child was an "environmentalist" in the now archaic sense that she took the side of nature in the age-old debate about whether nature or nurture determines human development. Whereas most Romantics believed in the unimpeded power of the human spirit to break worldly constraints and to create new selves and societies, "environmentalists" stressed the force of both natural and built surroundings to shape individual and social being.[29] Child was also an intersectional radical—a feminist, an abolitionist, an advocate of native rights, an opponent of capital punishment, and more. At the same time, she was a hard-working pragmatist with persistent financial challenges, who recognized that she must find common ground with readers who were suspicious of, and even hostile to, her ideas.[30]

In 1824, Child achieved sudden notoriety with the publication of her first novel, *Hobomok*, the story of Mary Conant, an Anglican woman who defies her domineering Puritan father and marries Hobomok, a Wampanoag man who had befriended the settlers of Plymouth. The book was denounced as "unnatural" and "revolting" by the *North American Review*, and a second edition was not printed.[31] Nevertheless, *Hobomok* established Child's reputation, and for the next decade she led a busy career as an historical novelist, writer of domestic advice books, and editor of a popular children's magazine, the *Juvenile Miscellany*. Then, the 1833 appearance of *An Appeal in Favor of That Class of Americans Called Africans*, one of the first abolitionist books published in the United States, angered many of her readers. Sales of her other titles fell, and her magazine folded.

In response to this reversal, Child immersed herself in research. She read hundreds of travel narratives and sacred texts in preparation for writing

The History of the Condition of Women in Various Ages and Nations (1835), an encyclopedic survey of the varieties of sexual oppression from antiquity to the present. Child placed *The History* with John Allen, who kept a bookshop at 11 School Street near Boston's Old South Meeting House. He was the publisher of *New Jerusalem Magazine*, the organ of the Swedenborgian New Church in the United States, and he also printed Bibles, almanacs, giftbooks, and the Popular Library series, which included miscellanies and travel narratives. Allen's publishing house was an important component of antebellum Boston's emergent culture of internationalism. His list, with its combination of unorthodox European theological tracts and sturdy moneymakers, was a natural home for Child's simultaneously domestic and transgressively global work. Allen released *The History of the Condition of Women in Various Ages and Nations* as the fourth and last title in Child's Ladies' Family Library series. The three previous volumes in the series, published by Carter and Hendee, were *The Biographies of Madame de Staël, and Madame Roland* (1832), *The Biographies of Lady Russell, and Madame Guyon* (1832), and *Good Wives* (1833). All three were steady sellers that required additional printings, but they were far outpaced by *The History*, which sold through four Boston editions (along with one in London) in eight years. Then, in 1845, Charles S. Francis published an abridged version of the book that sold through five printings by 1850. In short, *The History* held the steady attention of a wide audience in New England throughout the 15-year period during which the Transcendentalist movement flourished.

The contrast between the series title, Ladies' Family Library, and the volume title, *The History of the Condition of Women in Various Ages and Nations*, indexes the rhetorical situation that made the book so compelling. Child speaks to her audience in the cosmopolitan voice of a world traveler who has returned across the threshold into the domestic sphere. She delivers finely detailed experience of the entire globe, giving New England women vicarious access to a staggering variety of exotic places and peoples. *The History* marshals an encyclopedic spectacle of ethnographic information from hundreds of sacred texts, histories, travel narratives, expedition logs, journals, and other sources, focusing on the social position of women around the world from antiquity to the present. The book is geographically organized. Volume One, which is subtitled "Comprising the Women of Asia and Africa," starts in Palestine, then moves generally east, then returns to Egypt and continues south from there. Child begins by describing the lives of the women of Israel, then turns to (using her terms) the Babylonian, Lycean, Trojan, Greek, Syrian, Druse, Turkoman, Arabian, Bedouin, Afghan, Circassian,

Georgian, Armenian, Turkish, Persian, Hindu, Thibetian, Birmese, Cochin Chinese, Siamese, Malay, Chinese, Corean, and Tartar women; the Amazons, the Siberian, Javanese, and Sumatran women; the women of Borneo, Celebes, Amboyna, Bali, Timor, New Guinea, New Holland, Van Diemen's Land, the Philippine Islands, Loo Choo, Japan, the Fox Islands, and the Kurile Islands; and the Egyptian, Carthagenian, Moorish, African, Hottentot, and Dutch African women. Many of these categories, particularly "African women," are further subdivided by region and tribe. The second volume continues this world tour, recounting the lives of "the women of Europe, America and the South Sea Islands."

Child's tone throughout *The History* is directly factual as she describes courtship, marriage, and divorce customs; differential property laws; legal and political rights; divisions of labor; and religious and cultural traditions across cultures. She pays particular attention to clothing and housing arrangements for women, showing that there is wide variation across time and space in women's liberty to move and converse freely in public spaces. She remarks, for instance, that the "Persians seem to have been remarkable among the ancient nations for a savage jealousy of women, which led them to keep the objects of their love perpetually imprisoned and guarded. Their severity is spoken of as extraordinary, by Plutarch, and other authors, who wrote at a period when even the most enlightened nations allowed very little freedom to their women."[32] By connecting increases in women's physical autonomy to a Romantic narrative of global social evolution toward an enlightened present, Child implies that the antebellum period's intensifying restriction of women to the domestic sphere is both anachronistic and doomed to fail. She pays particular attention to the question of women's freedom of movement, not merely for its own sake, but especially because she sees a correlation with their intellectual and spiritual development. She remarks, for instance, that because

> Persian women are kept continually shut up in the harem ... all that is truly estimable in female character is neglected, as it ever must be where nothing like free and kind companionship exists between the sexes. A resident in Persia declares that the women are ignorant, and inconceivably gross in their ideas and conversation. Under such a system it could not be otherwise.[33]

Seemingly mundane facts about restrictive clothing, residential architecture, and customs governing public encounters between the sexes not only reflect, but explain, differences across cultures in women's ability to develop as individuals.

Child relates innumerable examples of "extraordinary customs," such as those of the "Nairs, on the coast of Malabar, [whose women] are usually married before they are ten years of age; but it would be deemed exceedingly indecorous for the husband to live with his wife, or even to visit her, except as an acquaintance. She lives with her mother, and prides herself on the number of her lovers, especially if they be Bramins or rajahs."[34] She also focuses especially closely on atypical incest taboos and especially harsh punishments for infidelity, as well as on examples of polygamy, widow sacrifice, and female infanticide. And she emphasizes examples of gender role reversal, noting, for instance, that in "the mountainous districts of China is a singular tribe called the Miao-Tse [where] the men wear ear-rings, and the women carry a sword. Both go barefoot, and climb the sharpest rocks with the swiftness of mountain goats."[35] Throughout the book, then, Child uses striking examples of cultural difference in order to destabilize her Euro-American readers' sense of the inevitability of the gender relations governing their own lives. In addition, Child regularly inserts anecdotal snapshots of women engaging in vigorous physical activity and sketches of women occupying positions of literary, judicial, political, military, intellectual, and religious power. "The women of Celebes" she writes, "are distinguished for virtue and modesty. They take an active part in business, and are frequently raised to the throne, though the government is elective. At public festivals they appear freely among men, and those in authority discuss affairs of state in their councils."[36] Similarly, in her description of the lives of Syrian women, she writes of the "celebrated queen ... Nitocris, wife of Nabonadius [who] was a woman of great endowments. While her voluptuous husband gave himself up to what the world calls pleasure, she managed the affairs of state with extraordinary judgment and sagacity. She was particularly famous for the canals and bridges which she caused to be made for the improvement of Babylon."[37] Child presents a composite picture of the startling diversity of gender relations across time and cultures, thereby demonstrating that the specific customs of New England in the 1830s were both local and contingent.[38] In order to make this point even clearer, when the time comes to describe the lives of her readers, the "Women of the United States," she gives them neither tonal emphasis nor pride of place within the panorama of the book. Their way of life, with all of its galling restrictions, is only one local variation on a global pattern of sexual oppression, a pattern within which there are nevertheless glimpses of possibility and hope.

Reflecting the character of her sources, Child reproduces many of the invidious binaries of Romantic historiography, but she does not do so uncritically. She reads human history, for instance, as a narrative of progress from savagery to civilization, and she tells exoticizing tales of "Mohammedan" sensuality, Aztec sacrificial rites, and so forth. At the same time, in a book that is remarkable for its almost complete lack of open editorial commentary, Child allows herself on several occasions to step from behind the veil of flat reportage to make damning analogies between various "savage" customs and the contemporary institution of American slavery:

> The Tchuktchi are among the wildest of the Siberians. They consider it wrong and disgraceful to rob or murder one of their own tribe; but such actions are regarded as honorable, and even glorious, when committed upon the members of any other tribe. This is a good commentary on Christian and enlightened nations, who consider it a great sin to make slaves of their own people, but regard the self-same action as perfectly justifiable toward persons of different complexion. If savage nations could write our history, how ridiculous they might make us appear by stating simple facts, without the varnish of sophistry with which we are accustomed to conceal them![39]

Similarly, while she sees the peoples that she describes as being arranged on a "scale of humanity" and she accepts the idea that "fine" features are more beautiful than "gross" ones, Child also scrupulously delinks "savagery" from complexion, asserting that "Africans consider our color quite as great a deformity as we regard theirs."[40] Moreover, she insists that both high civilizations and barbaric tribes have evolved on all continents, noting that "Egypt was the first nation that became civilized, and framed wise laws, by which they agreed to be governed. It had reached the height of its grandeur, and was beginning to decay, while nations which we call ancient were yet in their infancy."[41] Finally, rather than accepting racial categories as transhistorical and translocal, she offers what was until recently called an "environmentalist" explanation for variations in skin color across the globe:

> The ancient Jews were of the same dark complexion as the Arabs and Chinese; but their history furnishes a remarkable example of the influence of climate. They are dispersed all over the globe, and wheresoever they sojourn, those who marry among Gentile nations are cast out of the synagogue; therefore whatever changes have taken place in their color must have been

induced by climate and modes of life. There are now Jews of all complexions, from the light blonde of the Saxon to the deep brown of the Spaniard, and the mahogany hue of the Moors.[42]

The hierarchies of Romantic racism disintegrate here under the pressure of Child's attention to change over time in response to local environmental conditions.

Similarly, Child explains variations in the sexual division of labor across cultures as grounded in "modes of life": tribal, nomadic, agricultural, feudal, and so on. Speaking of rigid sexual hierarchy among the biblical Israelites, she explains:

> In those times, when the earth was thinly peopled, an increase of laborers was an increase of wealth; hence, physical strength, being the quality most needed, was most esteemed. To be the mother of a numerous family was the most honorable distinction of women; and the birth of a son was regarded as a far more fortunate event than the birth of a daughter. Under such circumstances, women were naturally considered in the light of property; and whoever wished for a wife must pay the parents for her.[43]

Child implies here that customs and ideas regarding gender relations evolve in response to the material pressures of a society's ways of organizing the production of food, shelter, fuel, fiber, and the other necessities of daily life. There were exceptions to this rule, to be sure. Child observes, for instance, that in ancient Greece, "certain women disguised themselves in male attire, and went to Academus to listen to the philosophy of Plato; and when this desire for knowledge began to prevail, it could not be long before it manifested itself in casting off the fetters prescribed by custom." By way of explaining what she represents as an anomaly, she remarks, "Individuals there were, as there ever will be, of both sexes, who were in advance of the people among whom they lived."[44] Child implies, though, that such individuals are rare and that their influence is often transitory, outmatched by the constraints of cultures whose internal hierarchies are central to their strategies for surviving in particular environments.

While Child's environmental feminism in *The History* may seem mostly implied, it comes into clearer relief when contrasted with the explicit idealism and essentialism of her friend Margaret Fuller's 1843 manifesto, "The Great Lawsuit. Man versus Men. Woman versus Women."[45] At the heart of "The Great Lawsuit" lies the universalizing claim that "petrified or oppressive institutions" have produced "a low materialist tendency" (i.e., a misguided obsession with financial gain) that distorts not just sexual relations, but human social relations as a whole. For Fuller, ideology is the driving force of sexism and other forms of oppression, so she looks forward to an organic revolution, in which exceptional individual women, by pursuing their own self-culture, will transform society by the force of their example: "the gain of creation consists always in the growth of individual minds, which live and aspire, as flowers bloom and birds sing, in the midst of morasses." Fuller envisions a day when such women will invalidate the doctrine of spheres, claiming "the intelligent freedom of the universe ... with God alone for their guide and their judge."[46] The key to transforming a sexist society, then, is to remove "artificial obstacles," the ideological, cultural, and legal conventions that restrict the development of women as individuals: "We would have every arbitrary barrier thrown down. We would have every path laid open to woman as freely as to man." The results would be not merely the liberation of individual women, but a revolutionary transformation of society as a whole, a new dawn of "ravishing harmony."[47]

As Child's remarks on the women of Greece show, she agreed with Fuller about the importance of public self-transformation. In a February 1842 letter to abolitionist Francis Shaw, she writes, "Now my own opinion is that the perfection of the *individual* is the sure way to regenerate the *mass*. I am to obey my highest instincts; and in no other way can I possibly do so much to bring discordant social relations in harmony."[48] On the other hand, Child also recognized that improvements in the condition of women were made possible by the historical process of ecosocial evolution. Moreover, she acknowledged that this process could not be viewed as a simplistic narrative of uninterrupted progress. For instance, while early agricultural societies may have restricted women to the physical reproduction of labor, the more complex class societies that developed with the accumulation of material surpluses sometimes offered women moments of fluidity in which they might seek to assert new kinds of freedom. Doing so could produce unpredictable results. On the one hand, liberty might lead to license. For instance, in "the earliest and best days of Rome, the first magistrates and generals of armies ploughed their own fields, and threshed their own grain. Integrity, industry, and simplicity, were the prevailing virtues of the times; and the character of women was, as it always must be, in accordance with that of men."[49] As long as Roman culture retained its "primitive simplicity, mothers nursed their own babes," she writes, "but as luxury increased, indolence and love of pleasure so far conquered maternal affection, that women of rank almost universally consigned their children to the care of female slaves."[50] On the other hand, Child suggests that the "early civilization" of Egypt "might be in part owing to the annual over-flowing of the Nile" during which families "were obliged to take shelter in houses raised on piles above the reach of the waters." During the annual flood season, "men and women, being thus placed in each other's society, naturally endeavored to please each other, and female influence produced its usual effect of softening the character, and rendering the manners more polished and agreeable." Not only were gender relations improved, but "from this union, music, poetry, and the fine arts would naturally flow, as the stream from its parent fountain."[51] For Child, then, changes in the condition of women were driven by material pressures specific to widely varying processes of social adaptation to local environmental conditions. By implication, as the antebellum Northeast rapidly transformed itself from a patchwork of agrarian villages into an integrated network of commercial cities and factory towns surrounded by consolidated agricultural districts, women were faced with a rare moment of opportunity during which they might claim new forms of power and freedom. Doing so, however, would require a clear understanding of the giant new cities taking shape in the region and especially an understanding of the new kinds of gender relations and new forms of sexual oppression that were developing within them.

THE GREEN CITY IN *LETTERS FROM NEW-YORK*

Child would pursue an opportunity to develop such an understanding when she decided to move alone to New York City a few years later. After the vicious reviews of her *Appeal*, Child and her husband David had dedi-cated themselves more and more completely to the growing movement for the abolition of slavery. In September 1835, Child wrote to her brother Convers Francis, "I believe that the world will be brought into a state of order through manifold revolutions." She encouraged him not to com-plain about having been "cast on these evil times," and she enjoined him instead "to rejoice that we have much to do as mediums in the regenera-tion of the world." A few months later, Child remarked on her own radi-calization in a letter to her mother: "my democracy increases with my years."[52] The "Woman Question" increasingly polarized the anti-slavery movement after the Grimke sisters came to prominence, and Child enthu-siastically defended the rights of women, not only to speak publicly against slavery, but to join and lead abolitionist organizations. Then, in 1838, both her activism and her writing stopped when she and David moved to Northampton in western Massachusetts, where they threw themselves into an attempt to solve their persistent financial problems by producing beet sugar, which they hoped to market as a product that was free of slave labor. For the next three years, the cosmopolitan and outspoken Child found herself isolated in an oppressively conservative rural village that was a favorite vacation spot for slaveholders. Writing to Abigail Kelley soon after moving to Northampton, Child dispels the idea that it was "a delightful rural retreat" where she had been "luxuriating in the beauties of Nature and Art." Instead, she complains that since arriving, she had "hungered and thirsted after the good, warm abolition-sympathy" and for "the din and activity of Boston."[53] She complains about the pastoral setting again in a December 1838 letter to Convers Francis: "If I were to choose my home, I certainly would not place it in the Valley of the Connecticut. It is true, the river is broad and clear, the hills majestic, and the whole aspect of outward nature is most lovely. But oh! the narrowness, the bigotry of man!"[54] Ironically, after three years in the stifling atmosphere of Northampton, she would discover the "beauties of Nature" in the biggest city in North America.

In April 1841 Child was offered the editorship of the *National Anti-Slavery Standard*, a weekly newspaper published in New York by the American Anti-Slavery Society. She accepted the position and moved to Manhattan a month later, leaving her husband David behind to manage the farm. From the beginning, she was determined to keep the *Standard* out of the bitter sectarian fights raging between the old Garrisonian and the new Liberty Party wings of the abolition movement.[55] Rather than engage in what she saw as ugly doctrinal wrangling between "ultra abolitionists," she set out to produce what she called "a good family anti-slavery newspaper" that would "gain the ear of the people at large."[56] So, in addition to publishing transcripts of abolitionist speeches, reports on relevant legislation, correspondence from abolitionist readers, and exposés of Southern cruelty, Child printed her own short essays in almost every issue of the newspaper from August 1841 until she resigned in May 1843. Many of these pieces address the reader directly as "you" and take on the tone of familiar correspondence. Child takes on a startlingly wide range of topics and experiences, and she experiments broadly with the generic possibilities of the familiar or personal essay.[57] However, she almost never writes openly about slavery or other social issues. Despite their seeming quietism, when these essays are read in context, surrounded as they were by explicitly abolitionist material, they can be recognized as reform texts that attempt to engage readers through indirect rhetorical means. They engage in a kind of "warfare without poison arrows, [which is] fought on the broad table land of high mountains, never descending into the narrow by-paths of personal controversy, or chasing [the] foe through the crooked lanes of policy."[58] Child's goal was to pursue social change by Transcendentalist means. As Bruce Mills puts it, she "uses New York realities as emblems for transcendent truth and, in doing so, seeks to reform old stereotypes and sterile customs."[59]

In August 1843, Child gathered 40 of her essays into a manuscript, which she edited intensively with help from her close friend and correspondent, Ellis Gray Loring. In a letter to Loring about the project, she wrote that she hoped to help readers "look candidly at anti-slavery principles" by "drawing them with the garland of imagination and taste."[60] Boston's eminent Transcendentalist printer, James Munroe, and his long-time New York collaborator, Charles S. Francis, recognized the value of the manuscript and rushed it into print as *Letters from New-York*. The first run of 1500 copies sold out by the end of the year, and the book went through ten more editions by the end of the decade.[61] The front matter of the first edition reflects Child's careful management of her authorial persona.[62] Beneath her byline on the title page, she is identified as the "author of *The Mother's Book, The Girl's Book, Philothea, History of Women*, etc." This sequence of titles emphasizes both her benign domesticity and her credentials as an eminent historian and novelist. The letters extend the transgressive persona she had established in *The History*, as she repeatedly insists on her own femininity, and then recounts one fearlessly transgressive excursion after another into the definitive masculine/capitalist space, Manhattan.[63]

In addition to its significance as an experiment in Transcendental reform rhetoric by a public woman, *Letters from New-York* is also historically important for the way that it challenged contemporaneous textual representations of New York and, by extension, of the modern city.[64] The 1820 U.S. census reported Boston's population as 43,298 and New York's as 123,706. Twenty years later, Boston's population had more than doubled to 93,383, while New York's had almost tripled to 312,710. On the eve of the Civil War, Boston's population had almost doubled again to 177,840, while New York's had reached a staggering 813,669. Both cities were chaotic spaces, built with only rudimentary planning. Shining commercial cores and elite neighborhoods were surrounded by alarmingly crowded slums whose residents lacked even the most basic sanitation and infrastructure. Not surprisingly, these disease- and disaster-ridden neighborhoods saw high rates of compensatory consumption and crimes of poverty. Explosively growing cities like these were something new in the New World, and they were the subject of intense curiosity and anxiety. Throughout the antebellum decades, cities and their dangers were the main subject of hundreds, perhaps thousands, of dime romances, factory novels, and lurid temperance tracts.[65] In them, Boston and its surrounding mill towns were portrayed as stiffly moralistic places, where a small number of fantastically wealthy and powerful families kept tight reins on workers and the poor, exercising an especially harsh kind of patrician authority over the displaced farm girls who mainly staffed their kitchens and factories. New York, on the other hand, was represented as a warren of depravity, a terrifying place where lechery, prostitution, robbery, fraud, and drunkenness ran rampant.

One of the earliest of these texts was Sarah Savage's *The Factory Girl* (1814), which tells the story of a young woman's struggles to maintain her dignity and piety in a mill town where the other workers are frivolous and immoral. Through the 1820s, there was a trickle of similar texts, but in the 1830s, and especially after the Panic of 1837, this trickle became a flood of increasingly histrionic books and chapbooks with titles like the following:

The Poor Rich Man and the Rich Poor Man (1836)
Charcoal Sketches; or, Scenes in a Metropolis (1838)
The Adventures of Harry Franco: A Tale of the Great Panic (1839)
The Two Defaulters; or, A Picture of the Times (1842)
The Elliott Family; or, The Trials of New-York Seamstresses (1845)
Helen Howard, or The Bankrupt and the Broker (1845)
The Belle of the Bowery; A Tale of New York City (1846)
Clarilda: or, The Female Pickpocket (1846)
Debtor and Creditor: A Tale of the Times (1848)
Sharps and Flats; or, The Perils of City Life (1850)
New York by Gas-Light; With Here and There a Streak of Sunshine (1850)

The most successful of the hundreds of antebellum exposés of urban life was George Lippard's seduction and murder story, *The Quaker City; or, the Monks of Monk-Hall* (1844), which was the best-selling novel in the United States before Harriet Beecher Stowe's *Uncle Tom's Cabin*. In almost all of these sensationalistic fictions, the modern commercial or industrial city is described as a filthy, crowded maze of streets in which decadent bankers, lawyers, merchants, and factory owners prowl for innocent, poverty-stricken young women who have left rural homes in search of work.

One especially lurid text was Harrison Gray Buchanan's *Asmodeus, or, the Iniquities of New York: A Complete Expose of the Crimes, Doings and Vices, Both in High and Low Life, Including the Life of a Model Artist* (1849). The frontispiece of the book is an engraving of four sleeping women, their breasts exposed and limbs entwined, draped in neoclassical robes and reclining on ornate couches. A single male arm reaches menacingly into frame from the top-right corner. The book opens with an introductory essay consisting of a series of extracts from a "Report of Arthur Tappan on the Magdalens of New York," accompanied by hyperbolic lamentations about the "TWENTY THOUSAND WOMEN in the City of New York who drive that most dreadful of all trades – the traffic of their virtue for gold."[66] This introduction serves as a frame for several short stories, assuring the reader that just as "the Spartans exhibited drunken helots before their children, to incite in their minds an abhorrence of the drunkenness, so we here present to the world the following pictures of sin and depravity, copied from the columns of the Police Gazette."[67] The first story concerns a poor woman, Jane Morrison, whose health has been ruined by sewing and who now poses as a "model artiste" in order to earn money. Even though those "who witnessed her personation of the Greek Slave, have avowed that her figure and her graceful ease surpassed that of the creation of the genius of Powers," she inexorably descends into poverty, vice, and consumption.[68] The second story tells of three sisters who are forced to exhibit themselves for money by their drunken mother. This is followed by tales of counterfeiters, gamblers, thieves, men who betray their mistresses, profligates who indulge in orgies of oyster eating, victims of homelessness and hunger wallowing in squalid tenements and brothels, a wife with two husbands, and so on. In one particularly overdrawn sketch, a starving young woman languishes behind an empty beer barrel in an abandoned brewery, sucking the blood from her own arm in order to stay alive.

Child's *Letters from New-York* cuts sharply against the image of the city that dominates antebellum popular literature.[69] Her book's first four letters reject existing conventions, establish an alternative mode of representing the city, propose an explanation of the worst features of the urban environment, and offer a solution to them. Recording her "first impression" of "this Great Babylon" in "Letter I," Child writes, "we arrived at dawn, amid fog and drizzling rain, the expiring lamps adding their smoke to the impure air, and [we saw] close beside us a boat called the 'Fairy Queen,' laden with dead hogs."[70] After quickly establishing an atmosphere of commercial grime with this vignette, Child widens her angle and gives a broad sketch of New York as a "hive" where "Wealth dozes on French couches, thrice piled, and canopied with damask, while Poverty camps on the dirty pavement, or sleeps off its wretchedness in the watch-house." The invidious contrast here between wealth and poverty would have been quite familiar to readers of contemporaneous descriptions of the city. So would Child's formulaic critique of modernity in her next paragraph: "In Wall-street, and elsewhere, Mammon, as usual, coolly calculates his chance of extracting a penny from war, pestilence, and famine; and Commerce, with her loaded drays, and jaded skeletons of horses, is busy as ever fulfilling the World's contract with the devil." The deliberate use of stock phrases and emblems here signals that Child intends to transcend the common discourse of temperance tracts and urban exposés, that she plans to widen her angle beyond such proverbial binaries as "magnificence and mud, finery and filth, diamonds and dirt."[71] She explains, "I deliberately mean to keep out of sight [the] disagreeables of New-York [since by] contemplating beauty, the character becomes beautiful."[72]

For Child, beauty often takes a quite specific form. She reports that, as she has grown accustomed to life in the city, "bloated disease, and black gutters, and pigs uglier than their ugly kind, no longer constitute the foreground in my picture of New-York. [Instead,] I have become more familiar with the pretty parks, dotted about here and there; with the shaded alcoves of the various public gardens; with blooming nooks, and 'sunny spots of greenery.'"[73] Most of the remainder of "Letter I" is devoted to a rhapsodic description of the Battery, which Child asserts "rivals our beautiful Boston Common."[74] She encourages the skeptical reader to "go there in the silence of midnight, to meet the breeze on your cheek, like the kiss of a friend; to hear the continual plashing of the sea, like the cool sound of oriental fountains ... to look on the ships in their dim and distant beauty, each containing within itself, a little world of human thought, and human passion."[75] Child concludes her visit to the Battery by asking her readers' indulgence for having made such a seemingly trivial literary excursion: "Therefore blame me not, if I turn wearily aside from the dusty road of reforming duty, to gather flowers in sheltered nooks, or play with gems in hidden grottoes. The Practical has striven hard to suffocate the Ideal within me; but it is immortal, and cannot die. It needs but a glance of

Beauty from earth or sky, and it starts into blooming life, like the aloe touched by fairy wand."[76] In other words, Child suggests by the example of her own experience that intellectual and emotional health requires denizens of the city to reorient themselves from time to time by making contact with nature where they can.[77]

"Letter II" returns to the city's "sunny spots of greenery," praising the neatness of the Washington Parade Ground and Union Park, but remarking that "St. John's Park, though not without pretensions to beauty, never strikes my eye agreeably, because it is shut up from the people; the key being kept by a few genteel families in the vicinity." Suddenly, Child's cheerful talk about parks takes on a new tone as she notes that she is "an enemy to monopolies" and wishes for "all Heaven's good gifts to man to be as free as the wind, and as universal as the sunshine .... Let science, literature, music, flowers, all things that tend to cultivate the intellect, or humanize the heart, be open to 'Tom, Dick, and Harry' .... In all these things, the refined should think of what they can *impart,* not of what they can *receive.*"[78] Child represents parks here as a necessary setting for the kind of self-culture that Boston's Transcendentalists, following the paternalistic William Ellery Channing, proposed as the best means of addressing the poverty and criminality of the new urban working class. "As for the vicious," she writes, "they excite in me more of compassion than of dislike .... God forbid that I should wish to exclude them from the healthful breeze, and the shaded promenade."[79] This thought reminds Child that "in this vast emporium of poverty and crime, there are, *morally* speaking, some flowery nooks, and 'sunny spots of greenery.'" And she dedicates the remainder of this letter to an example of moral greenery, the Washington Temperance Society. Child praises this organization of former alcoholics because, rather than condemn the drinker, they "gently [draw] him within the golden circle of human brotherhood." And she remarks that the group "is one among several powerful agencies now at work, to teach society that it *makes its own criminals,* and then, at prodigious loss of time, money, and morals, punishes its own work."[80] This second letter, then, transforms green space from a personal convenience into a matter of public policy. Parks are no longer just pleasant retreats for individual recreation; instead, they take their place in a panoply of antebellum strategies for reform, from campaigns for literacy and spiritual awakening to efforts to eradicate prostitution and alcoholism, all of which were designed to address the social pathologies that afflicted workers and the poor in the new city.

"Letter III" suggests that urban misery is directly caused by the eradication of green space. Child begins with a meditation on a "brick wall staring in at my chamber window," then glowingly describes "a little, little patch of garden" and "two beautiful young trees," an ailanthus intertwined with a catalpa, that are also visible from her room. To readers who might wonder at her enthusiasm for these seeming trivialities, she writes, "You too, would worship two little trees and a sunflower, if you had gone with me to the neighbourhood of the Five Points the other day." Five Points was a notorious Manhattan slum; Child describes it as "an open tomb" where "you will see nearly every form of human misery, every sign of human degradation. The leer of the licentious, the dull sensualism of the drunkard, the sly glance of the thief – oh, it made my heart ache for many a day." Child takes special notice of "multitudes of children – of little *girls*," leaving the reader to imagine the varieties of degradation to which they are subject.[81] Then, in what seems at first to be a wrenching change of subject, she remarks: "It is said a spacious pond of sweet, soft water once occupied the place where Five Points stands. It might have furnished half the city with the purifying element; but it was filled up at incredible expense – a million loads of earth being thrown in, before perceivable progress was made. Now, they have to supply the city with water from a distance, by the prodigious expense of the Croton Water Works." This interjection is more than an ironic historical anecdote; for Child, it "is a good illustration of the policy of society towards crime. Thus does it choke up nature, and then seek to protect itself from the result, by the incalculable expense of bolts, bars, the gallows, watch-houses, police courts, [and] constables."[82] The city, where nature has been almost entirely displaced by a built environment of maximum utility, concentrates the poor, focusing their misery on each other and giving rise to new kinds and intensities of crime.

After anatomizing and diagnosing New York in her first three letters, Child opens "Letter IV" by acknowledging the "great privilege" that the city enjoys "in facility and cheapness of communication with many beautiful places in the vicinity. For six cents one can exchange the hot and dusty city, for Staten Island, Jersey, or Hoboken; three cents will convey you to Brooklyn, and twelve and a half cents pays for a most beautiful sail of ten miles, to Fort Lee." By way of sampling the city's nearby recreational resources, Child visits the village of Hoboken, where a "small open glade, with natural groves in the rear, and the broad river at its foot, bears the imposing name of Elysian Fields."[83] But she finds, to her dismay, that "the city intrudes her vices into this beautiful sanctuary of nature," for there is a "public house" in the Elysian Fields, "with its bar room, and bowling alley, a place of resort for the idle and profligate."[84] Next to the public house, she encounters "two tents of Indians," a "remnant of the Penobscot tribe" who are selling baskets to survive. Then, on a second visit to Hoboken, Child walks along the river to Weehawken, where she is reminded even more forcibly of the city she has left temporarily, for she happens upon "the supposed scene of the Mary Rogers' tragedy," an 1841 murder case that shocked New York, especially because it was rumored that Rogers died during a botched abortion. She writes, "I could not forget that the quiet lovely path we were treading was near to the city, with its thousand hells, and frightfully easy of access."[85] Child concludes her fourth letter, then, from the vantage point of Hoboken, where she faces the disturbing prospect that New York threatens to infect nature itself with the endemic diseases of the modern city.

While Child does not return to sustained discussion of green space in the remainder of *Letters from New-York*, the idea of nature as both material environment and spiritual home grounds her subsequent meditations on topics as wonderfully diverse as New York's antiquities, musical harmony, animal magnetism, the morality of war, the spirituality of flowers, the irrationality of imprisonment, Croton Aqueduct, capital punishment, women's rights, and lightning. Heather Roberts argues that Child creates an "emotionally engaged and engaging narrative persona" in these letters, a "'public heart' [that] models a sympathetic alternative to the detached voyeurism of the traditional urban flaneur and ultimately gestures toward a new, more inclusive and multicultural 'Christian cosmopolitanism.'"[86] For instance, when she passes Blackwell's Island (now Roosevelt Island) with its "Lunatic Asylum" and penitentiary, she embodies reforming sympathy by asking whether the "morally and the intellectually insane [should] not both be treated with great tenderness?" In answer to this rhetorical question, Child emphasizes the therapeutic value of green space by observing that there has been "kindness evinced in the location chosen" for the asylum, "for if free breezes, beautiful expanse of water, quiet, rural scenery, and 'the blue sky that bends o'er all,' can 'minister to the mind diseased,' then surely these forlorn outcasts of society may here find God's best physicians for their shattered nerves."[87] Child's prescription for the residents of Blackwell's Island is clearly meant for the inmates of Manhattan as well. After all, alienation from nature lies at the heart of Child's diagnosis of what she calls the "age of Commerce."[88] In Letter XVIII, she contrasts New York's streets with the rural landscapes they have replaced:

Maiden-lane is now one of the busiest of commercial streets; the sky shut
out with bricks and mortar; gutters on either side, black as the ancients
imagined the rivers of hell; thronged with sailors and draymen; and redolent
of all wharf-like smells. Its name, significant of innocence and youthful
beauty, was given in the olden time, when a clear, sparkling rivulet here
flowed from an abundant spring, and the young Dutch girls went and came
with baskets on their heads, to wash and bleach linen in the flowing stream,
and on the verdant grass.[89]

Not only does Child nostalgically mourn an idyllic rural past in which
women worked together in nature, but she also explicitly anathematizes
the present, for "the spirit of Trade ... reigns triumphant, not only on
'Change, but in our halls of legislation, and even in our churches.'"[90] She
takes sardonic pleasure in describing "an ugly, angular building" as "a cari-
cature likeness of the nineteenth century." It "stands at the corner of
Division-street, protruding its sharp corners into the midst of things,
determined that all the world shall see it, whether it will or no, and covered with signs from cellar to garret, to blazen forth *all* it contains."[91] In
this relentlessly monetized present, homeless children beg on the side-
walks outside shops that display gleaming "vases of gold and silver" in
their windows.[92]

Such an ugly, instrumental cityscape cannot fail to affect its inhabitants.
On New York's streets, the "busy throng, passing and repassing, fetter
freedom, while they offer no sympathy [and] the loneliness of the soul is
deeper, and far more restless, than in the solitude of the mighty forest."[93]
By way of demonstrating the emotional pressures of life on the capitalist
streets of New York, Child devotes an entire letter to Macdonald Clarke,
a poet who was universally regarded as insane because all "that he had –
money, watch, rings, were given to forlorn street wanderers." Macdonald
is a modern saint driven to madness by the "continual torture of this great
Babel of misery and crime."[94] There are even times when Child seems to
follow Macdonald into despair: "A few days since, cities seemed to me
such hateful places, that I deemed it the greatest of hardships to be pent
up therein. As usual, the outward grew more and more detestable, as it
reflected the restlessness of the inward."[95]

Each time, though, that Child begins to descend into hopelessness, she
recovers when she encounters a remnant of nature surviving in the city,
such as a "multitude of doves ... wheeling in graceful circles, their white
wings and breasts glittering in the sunshine" on Broadway. The sight of
these birds alters her perspective and brings on a revelation: "The fault,"

she writes, "was in my own spirit rather than in the streets of New-York .... Had my soul been at one with Nature and with God, I should not have seen *only* misery and vice in my city rambles."[96] In Letter XXIV, Child renders this idea in the form of an emblem. Once again her "spirit is at war with the outward environment" but a friend, who has "faith in Nature's healing power," encourages her to go for a walk. While "yet amid the rattle and glare of the city, close by the iron railway, I saw a very little, ragged child stooping over a small patch of stinted, dusty grass. She rose up with a broad smile over her hot face, for she had found a white clover!"[97]

Unfortunately, one of the primary symptoms of New York's illness is that its sufferers feel compelled to destroy the very thing that can cure them. During a visit to Brooklyn Heights, Child sees fresh evidence that the city now threatens nature itself: "A few years ago, these salubrious heights might have been purchased by the city at a very low price, and converted into a promenade of beauty unrivalled throughout the world; but speculators have now laid hands on them, and they are digging them away to make room for stores, with convenient landings from the river."[98] By way of contrast, she devotes the second half of this letter to praising the nearby establishment of Greenwood Cemetery, whose proprietors "have the good taste to leave the grounds as nearly as possible in a state of nature."[99] Child quietly sets up similar contrasts between examples of destructive exploitation and of responsible urban land stewardship whenever the opportunity arises.

After all, despite the clarity of her vision, Child remains an optimist. This is partly a matter of principle, since she believes that in "this working-day world, where the bravest have need of all their buoyancy and strength, it is sinful to add our sorrows to the common load."[100] But she also remains convinced by narratives of historical progress: "After all, this nineteenth century, with all its turmoil and clatter, has some lovely features about it. If evil spreads with unexampled rapidity, good is abroad, too, with miraculous and omnipresent activity. Unless we are struck by the tail of a comet, or swallowed by the sun meanwhile, we shall certainly get the world right side up, by and by."[101] In other words, at the same time that Child laments the inhumanity of the modern commercial city, she believes that capitalism will, perhaps despite itself, produce positive changes in the world by bringing diverse people into contact, by creating opportunities for association and sympathy. She looks forward to a day when "Commerce will ... fulfill its highest mission, and encircle the world in a golden band of brotherhood."[102] In many of the letters, she explores the "infinite varieties of character" in this place of "rapid fluctuation" and "never-ceasing change," where one encounters on a daily basis "the Catholic kneeling before the Cross, the Mohammedan bowing to the East, the Jew veiled before the ark of the testimony, the Baptist walking into the water, the Quaker keeping his head covered in the presence of dignitaries and solemnities of all sorts, and the Mormon quoting from the Golden Book which he has never seen."[103] As Travis Foster argues, Child takes "sympathy into settings ... where the borders between separate races and between self and other cannot be neatly maintained, and she praises states of being in which racial subjectivity breaks down."[104] New York's diversity is more than ethnic and religious, though; because of the city's status as a hub of global commerce, the "enterprising, the curious, the reckless, and the criminal, flock hither from all quarters of the world, as to a common centre." And in such a place of constant change, people live "with somewhat of that wild license which prevails in times of pestilence. Life is a reckless game, and death is a business transaction. Warehouses of ready-made coffins, stand beside warehouses of ready-made clothing, and the shroud is sold with spangled opera-dresses." Child remarks that this sublime spectacle of a radically diverse humanity in frantic motion "sometimes forced upon me, for a few moments, an appalling night-mare sensation of vanishing identity; as if I were but an unknown, unnoticed, and unseparated drop in the great ocean of human existence."[105] Nevertheless, throughout the book, Child celebrates the city's internationalism and human variety with genuine verve. One letter describes traditional "Highland exercises" hosted by a Scottish benevolent society. Another describes a visit to "the Jewish Synagogue in Crosby-street, to witness the Festival of the New Year."[106] A third extols the powerful preaching of Julia Pell, an African American Methodist who seemed to "poise herself on unseen wings, above the wondering congregation."[107]

Letter XVI, with its description of a "great fire," may well be the most powerful and thought-provoking in the book: "It began at the corner of Chrystie-street, not far from our dwelling; and the blazing shingles that came flying through the air, like a storm in the infernal regions, soon kindled our roof." Soon, "the block opposite us was one sheet of fire." Child flees through streets "filled with a dense mass of living beings" and experiences a moment of disorientation. "Nothing surprised me," she writes, "so much as the rapidity of the work of destruction. At three o'clock in the afternoon, there stood before us a close neighborhood of houses, inhabited by those whose faces were familiar, though their names were mostly unknown; at five the whole was a pile of smoking ruins."[108] Faced with this wholesale destruction, Child feels "a deeper sympathy" for "Jane Plato, the coloured woman" whose "neatly-kept garden" has burned, along with the intertwined ailanthus and catalpa trees outside her chamber window. Child believed that slavery's cruelest punishment was to deny the slave a home and that a free black's most important requirement was land on which to make one.[109] In *Letters from New-York*, Jane Plato's garden serves as a homeplace in this expanded sense not only for a representative free black woman, but also for Child; it is a correlative of the letters themselves, which she describes as "sheltered nooks" where she rests from her travels on the "dusty road of reform duty."[110] So, while she acknowledges that her grief over the loss of trees will be regarded as perverse by the "utilitarian and the moralist," since the garden would "not be worth much in dollars and cents," Child nevertheless stresses that "it was to [Plato] the endeared companion of many a pleasant hour" of recreation after "her daily toil."[111] After suggesting that the burning of mere houses is no real tragedy as compared with the loss of a treasured green homeplace, Child remarks that the "great fire, like all calamities, public or private, has its bright side." After all, she writes, a "portion of New-York, and that not a small one, is for once thoroughly cleaned; a wide space is opened for our vision, and the free passage of the air." The fire, "so fearful in its beauty,"[112] has not only revealed the fragility of the seemingly permanent structures of the capitalist city, but also created an opportunity; it has cleared a space on which Jane Plato's garden might be replanted on a larger scale.[113]

In 1844, there was a changing of the guard. After retiring from her position at the *National Anti-Slavery Standard*, Child spent the year editing *Letters from New-York: Second Series*. In December, Margaret Fuller moved to New York to work for Horace Greeley's *New York Tribune*, where she would follow Child's lead as the city's foremost intellectual woman, commenting weekly on life and literature. Fuller reviewed *Letters from New-York: Second Series* in May 1845, immediately after it appeared in stores. She praises the book for its "familiar freedom" and for its "true and companionable insight into the purposes, no less than the symptoms, of our life." She also clearly states her admiration for the gender nonconforming persona that Child had developed, remarking that Child's "acquaintance with common life is large, and various, such as can be won only by powers of ardent sympathy, balanced by a love of justice." Finally, she recommends the book heartily to readers who seek "well defined and sympathetic narratives of events that lie all around us, but which few have eyes to see or hearts to understand."[114] While Fuller was at the *Tribune*, she focused almost entirely on literary criticism, though she did reprise

Child's public-feminist role in a few accounts of excursions, such as "Our City Charities," which describes her "Visit to Bellevue Alms House, to the Farm School, the Asylum for the Insane, and Penitentiary on Blackwell's Island."[115] In the *Tribune* essays, Fuller finds opportunities to comment on the full range of reform questions, from suffrage, to women's rights and roles, educational and penal reform, and the abolition of slavery. However, she does not pursue Child's suggestion that a more natural city could cure the worst ills of modernity. Instead, it was the poet William Cullen Bryant who led New York's literary community in a decade-long campaign for green space that culminated in the 1857 ground-breaking for Central Park. Frederick Law Olmsted, Central Park's self-identified "Socialist Democrat" designer, articulated a social engineer's vision of the green city. He argued that the city needed "institutions that shall more directly *assist* the poor and degraded to elevate themselves."[116] Elevation of the working classes required spaces where people of all classes could mix, free of the pressures of economic competition. A large park that reproduced the pastoral "beauty of the fields, the meadows, the prairie, of the green pastures, and the still waters" would give park visitors "tranquility and rest to the mind."[117] In such a setting they could participate in what Olmsted called "gregarious receptive recreation," enjoying "the greatest possible contrast with the restraining and confining conditions of the town, those conditions which compel us to walk circumspectly, watchfully, jealously, which compel us to look closely upon others without sympathy."[118] Child's home city, Boston, would follow New York's lead two decades later, hiring Olmsted to supervise construction of its Emerald Necklace. In the ensuing decades, Olmsted and his sons would design hundreds of city parks and college campuses across the United States, giving ubiquitous physical form to Child's Transcendentalist critique of capitalist modernity and her environmental feminist prescription for its cure. Many of these urban green spaces are now favorite locations for intersectional rallies and demonstrations that demand wholesale ecosocial change and perform the hopeful *topos* of the green city.

NOTES

1. Women's March.
2. Vagianos and Dahlen. See Kings for a consideration of the evolving ethic framework of intersectional ecofeminism.
3. There is a rich and extensive tradition of ecofeminist theory, often rooted in literary studies, whose leading arguments have manifested in the street

politics of the intersectional movement described here. A few starting points are Gaard, *Critical Ecofeminism*; Griffin, *Woman and Nature*; Haraway, *Simians, Cyborgs, and Women*; Mellor, *Breaking the Boundaries* and *Feminism and Ecology*; Merchant, *Radical Ecology*; and Mies and Shiva, *Ecofeminism*.

4. Judith Butler, 10.
5. Wollstonecraft, *A Vindication*, 27.
6. Wollstonecraft, *A Vindication*, 39.
7. Wollstonecraft, *A Vindication*, 57.
8. Wollstonecraft, *A Vindication*, 41.
9. Wollstonecraft, *A Vindication*, 17.
10. Wollstonecraft, *A Vindication*, 18.
11. Wollstonecraft, *A Vindication*, 340.
12. Wollstonecraft, *A Vindication*, 43.
13. Wollstonecraft, *A Vindication*, 51 and 68.
14. Wollstonecraft, *A Vindication*, 52 and 72–73. Charlene Avallone offers an alternate reading of this passage, asking, "Does [Wollstonecraft] so much cede independence as locate any claim to male superiority in brute physical strength, thus rejecting any rational claim?" (personal correspondence, n.p.).
15. Wollstonecraft, *A Vindication*, 340.
16. Wollstonecraft, *A Vindication*, 22.
17. Wollstonecraft, *A Vindication*, 71.
18. Wollstonecraft, *A Vindication*, 77.
19. Wollstonecraft, *A Vindication*, 152.
20. Wollstonecraft, *A Vindication*, 81.
21. Wollstonecraft, *A Vindication*, 275.
22. Wollstonecraft, *A Vindication*, 294.
23. Wollstonecraft, *A Vindication*, 79.
24. Wollstonecraft, *A Vindication*, 111–112.
25. Wollstonecraft, *A Vindication*, 66.
26. Wollstonecraft, *A Vindication*, 83.
27. Wach describes Wollstonecraft's strong influence on Margaret Fuller and Caroline Dall (3–35).
28. Child, *Mother's Book*, 59.
29. The primary sense of the term "environmentalist" is still "a person who believes in or promotes the idea of the primary influence of environment … on development." While quite recent examples of this usage do exist, it is rapidly being displaced by the newer sense: "a person who is concerned with the preservation of the environment, especially from damage caused by human influence" (*Oxford English Dictionary*). In order to avoid confusion, I will use the term "materialist" to refer to the position in the nature-versus-nurture debate. "Materialism" has the benefit of

highlighting parallels to contemporary theoretical debates within literary and cultural studies. However, I will frequently take advantage of the overlap between the two senses of the word "environmentalist" in order to think through the consequences for contemporary environmentalism of Romantic environmentalism. After all, if we see ourselves as powerfully shaped by our natural and built environments, we have a strong motivation to ensure that both shape us in ways that are healthy.

30. The best biographies of Child are Karcher, *First Woman*, and Mills, *Cultural Reformations*.
31. Review of *Hobomok*, 263.
32. Child, *The History*, v. 1, 72.
33. Child, *The History*, v. 1, 77.
34. Child, *The History*, v. 1, 109.
35. Child, *The History*, v. 1, 160.
36. Child, *The History*, v. 1, 201.
37. Child, *The History*, v. 1, 26.
38. Karcher, *First Woman*, 225.
39. Child, *The History*, v. 1, 180–181.
40. Child, *The History*, v. 1, 292 and 262.
41. Child, *The History*, v. 1, 216.
42. Child, *The History*, v. 1, 23.
43. Child, *The History*, v. 1, 2.
44. Child, *The History*, v. 2, 6.
45. During the winter of 1839, Child attended Fuller's conversations at Elizabeth Peabody's West Street bookshop. The two renewed their friendship and visited each other frequently when Fuller lived in New York from December 1844 to August 1846. Fuller praised *Letters from New-York* in the *Dial* and Child read *Woman in the Nineteenth Century* in page proofs. Child's work at the *Standard* set an important precedent for the role Fuller played at Horace Greeley's *New York Tribune*, where she moved quickly out of her initial position as literary critic to write about the radical politics and chaotic streetscapes of the world's most advanced commercial city. In "The Great Lawsuit," Fuller does gesture briefly in the direction of a comparative analysis of sexism when she observes sardonically that "acknowledged slavery" is the only place where "women are on a par with men," since each "is a work-tool, an article of property–no more!" (24). For accounts of the Child-Fuller friendship, see Karcher, "Margaret Fuller and Lydia Maria Child," and Baer, "Mrs. Child and Miss Fuller."
46. Fuller, "Great Lawsuit," 23.
47. Fuller, "Great Lawsuit," 14.
48. Child, *Selected Letters*, 161.

49. Child, *The History*, v. 2, 35.
50. Child, *The History*, v. 2, 47.
51. Child, *The History*, v. 1, 216–217.
52. Child, *Selected Letters*, 39 and 47.
53. Child, *Selected Letters*, 89.
54. Child, *Selected Letters*, 102.
55. Karcher, *First Woman*, 267–270.
56. Child, *Selected Letters*, 175. While in conservative Northampton, Child became increasingly careful in choosing strategies of argument: "I often attack bigotry with 'a troop of horse shod with felt'; that is, I try to *enter* the wedge of general principles, letting inferences unfold themselves gradually" (*Selected Letters*, 109). In addition to acknowledging the power of Socratic indirection, Child remarked on the value of allowing exemplary acts to speak for themselves. In a September 1839 letter to the infamously pugilistic rhetorician William Lloyd Garrison, Child recounts a conversation with the Grimke sisters: "They urged me to say and do more about woman's rights, nay, at times, they gently rebuked me for my want of zeal. I replied, 'It is best not to *talk* about our rights, but simply go forward and *do* whatever we deem a duty'" (*Selected Letters*, 123). Three months later, she expanded upon this idea in a letter to her sister-in-law, Lydia Bigelow Child; women, she argued, should "*do* whatsoever seems to us right, with a modest freedom, as if none would question. If duty calls us into unusual modes of action, the world will best become accustomed to it and prepared for it, by simply *seeing* it done" (*Selected Letters*, 126).
57. Tingley, 56.
58. Child, *Letters from New-York*, 52.
59. Mills, *Cultural Reformations*, 75. An epigraph from Samuel Tayler Coleridge's "Dejection: An Ode" appears on the title page of the first edition of *Letters from New-York*:

> We receive but what we give,
> And in our life alone does Nature live;
> Ours is her wedding-garment, ours her shroud!
> And would we aught behold, of higher worth,
> Than that inanimate cold world allowed
> To the poor loveless ever-anxious crowd,
> Ah! from the soul itself must issue forth
> A light, a glory, a fair luminous cloud
> Enveloping the Earth–
> And from the soul itself must there be sent
> A sweet and potent voice, of its own birth,
> Of all sweet sounds the life and element!

60. Quoted in Karcher, *First Woman*, 301–302.

61. A sequel, *Letters from New-York, Second Series*, which appeared in 1845, consisted of letters published mainly in the *Boston Courier*. This volume too sold through several editions by 1850.

62. In "'My Twin Sister': George Sand, Lydia Maria Child, and the Epistolary Journalistic Essay," Charlene Avallone identifies an important transatlantic model for Child's transgressive authorial persona and practice.

63. See Nord on gender and walking in public in nineteenth-century English literature. Child dedicated *Letters from New-York* to John Hopper, the young lawyer who had been "as a brother" to her "in a city of strangers" during the two years she edited the *Standard*. She acknowledges in her dedication that "most of the scenes mentioned in these Letters" they had "visited together" (*Letters from New-York*, 3). However, Hopper remains invisible in the letters themselves.

64. Karcher, *First Woman*, 302.

65. Very little has been written about this archive of ephemera. For a survey of representations of the city in canonical antebellum American literature, see Stout.

66. Buchanan, 8.

67. Buchanan, 70.

68. Buchanan, 21.

69. Roberts, 756–757.

70. Child, *Letters from New-York*, 9.

71. Child, *Letters from New-York*, 9.

72. Child, *Letters from New-York*, 15.

73. In Samuel Taylor Coleridge's "Kubla Khan," the "sunny spots of greenery" in the palace gardens contrast with a "deep romantic chasm" that is "haunted/By woman wailing for her demon lover" (102–104).

74. Child, *Letters from New-York*, 10.

75. Child, *Letters from New-York*, 11.

76. Child, *Letters from New-York*, 12.

77. At the same time that "Letter I" establishes a strong thread of discussion about the importance of green space, it also explains the Transcendental-imagist method that Child will adopt most in the rest of the book: "There *was* a time when all these things would have passed by me, like the flitting figures of the magic lantern, or the changing scenery of a theatre, sufficient for the amusement of an hour. But now, I have lost the power of looking on the surface. Every thing seems to me to come from the Infinite, to be filled with the Infinite, to be tending toward the Infinite" (*Letters from New-York*, 10). Thus, Child writes, "Do I see crowds of men hastening to extinguish a fire? I see not merely uncouth garbs, and fantastic, flickering lights, of lurid hue, like a tramping troop of gnomes, but

straightway my mind is filled with thoughts about mutual helpfulness, human sympathy, the common bond of brotherhood, and the mysteriously deep foundations on which society rest; or rather, on which it now reels and totters" (*Letters from New-York*, 10). More often, though, Child does not record the thoughts that come to her when she looks below the surface; instead, she renders the surfaces and allows her readers to look into the Infinite for themselves. *Letters from New-York*, then, with its seemingly unconnected anecdotes and impressionistic descriptions, should be read then in much the same way as *The History of the Condition of Women*, as "a troop of horse shod with felt" that makes a reform argument by indirect means.

78. Child, *Letters from New-York*, 12.
79. Child, *Letters from New-York*, 12–13.
80. Child, *Letters from New-York*, 13.
81. Child, *Letters from New-York*, 17.
82. Child, *Letters from New-York*, 18.
83. Child, *Letters from New-York*, 18.
84. Child, *Letters from New-York*, 18–19.
85. Child, *Letters from New-York*, 21.
86. Roberts, 751–752.
87. Child, *Letters from New-York*, 43. See also the description of the quarantine ground on Staten Island and the Sailor's Snug Harbor in Letter XXI (91–94).
88. Child, *Letters from New-York*, 34.
89. Child, *Letters from New-York*, 78.
90. Child, *Letters from New-York*, 34. Heather Roberts notes that Child frequently "uses the city's physical structures, and in particular, its ubiquitous walls, to embody its materialism" (759).
91. Child, *Letters from New-York*, 56.
92. Child, *Letters from New-York*, 59.
93. Child, *Letters from New-York*, 59.
94. Child, *Letters from New-York*, 67.
95. Child, *Letters from New-York*, 102.
96. Child, *Letters from New-York*, 74.
97. Child, *Letters from New-York*, 102–103.
98. Child, *Letters from New-York*, 30–31.
99. Child, *Letters from New-York*, 32.
100. Child, *Letters from New-York*, 73.
101. Child, *Letters from New-York*, 93.
102. Child, *Letters from New-York*, 34.
103. Child, *Letters from New-York*, 43.
104. Foster, 3.

105. Child, *Letters from New-York*, 44.
106. Child, *Letters from New-York*, 22 and 24.
107. Child, *Letters from New-York*, 48.
108. Child, *Letters from New-York*, 70.
109. Melissa Fiesta notes argues that "realization of literal homeplaces through actual property ownership allows the freed to create their own rhetorical homeplaces for collective social agency" (271).
110. Child, *Letters from New-York*, 12.
111. Child, *Letters from New-York*, 70–71.
112. Child, *Letters from New-York*, 73.
113. Rosenthal comments on Child's use of gardens as figurative spaces of black women's resistance grounded in "floral counterdiscourse" in *A Romance of the Republic* (1867). See also Pratt, "Rebuilding Babylon."
114. Fuller, *Critic*, 119–120.
115. Fuller, *Critic*, 98–104.
116. Olmsted, *Papers*, v. 2, 234.
117. Olmsted, *Public Parks*, 23.
118. Olmsted, *Public Parks*, 21–22.

BIBLIOGRAPHY

Avallone, Charlene. 2012. My Twin Sister': George Sand, Lydia Maria Child, and the Epistolary Journalistic Essay. *George Sand Studies* 31: 31–49.
———. 2018. Personal Correspondence, 15 August.
Baer, Helene G. 1953. Mrs. Child and Miss Fuller. *New England Quarterly: A Historical Review of New England Life and Letters* 26 (2): 249–255.
Buchanan, Harrison Gray. 1848. *Asmodeus, or, the Iniquities of New York: A Complete Expose of the Crimes, Doings and Vices, Both in High and Low Life, Including the Life of a Model Artist.* New York: Garrett and Company.
Butler, Judith. 2015. *Notes Toward a Performative Theory of Assembly.* Cambridge, MA: Harvard University Press.
Child, Lydia Maria. 1824. *Hobomok, a Tale of Early Times.* Boston: Cummings, Hilliard, and Co.
———. 1831. *The Mother's Book.* Boston: Carter, Hendee, and Babcock. *Internet Archive.*
———. 1835. *The History and Condition of Women in Various Ages and Nations.* Boston: John Allen & Co. *Internet Archive.*
———. 1843. *Letters from New-York.* New York: Charles S. Francis & Co. *Internet Archive.*
———. 1982. In *Selected Letters, 1817–1880*, ed. Milton Meltzer and Patricia G. Holland. Amherst: University of Massachusetts Press.

Coleridge, Samuel Taylor. 1986. In *The Oxford Authors: Samuel Taylor Coleridge*, ed. H.J. Jackson. Oxford: Oxford University Press.

Fiesta, Melissa. 2006. Homeplaces in Lydia Maria Child's Abolitionist Rhetoric, 1833–1879. *Rhetoric Review* 25 (3): 260–274.

Foster, Travis M. 2010. Grotesque Sympathy: Lydia Maria Child, White Reform, and the Embodiment of Urban Space. *ESQ: A Journal of the American Renaissance* 56 (1): 1–32. *Project Muse*.

Fuller, Margaret. 1843. The Great Lawsuit. Man Versus Men. Woman Versus Women. *Dial* 4 (1): 1–47. GoogleBooks.

———. 2000. In *Critic: Writings from the New-York Tribune, 1844–1846*, ed. Judith Mattson Bean and Joel Myerson. New York: Columbia University Press.

Gaard, Greta. 2017. *Critical Ecofeminism*. Lanham: Lexington Books.

Griffin, Susan. 1980. *Woman and Nature: The Roaring Inside Her*. New York: Harper & Row.

Haraway, Donna. 1991. *Simians, Cyborgs, and Women: The Reinvention of Nature*. New York: Routledge.

Karcher, Carolyn. 1994. *The First Woman in the Republic: A Cultural Biography of Lydia Maria Child*. Durham: Duke University Press.

———. 2000. Margaret Fuller and Lydia Maria Child: Intersecting Careers, Reciprocal Influences. In *Margaret Fuller's Cultural Critique: Her Age and Legacy*, ed. Fritz Fleischmann, 75–89. New York: Peter Lang.

Kings, A.E. 2017. Intersectionality and the Changing Face of Ecofeminism. *Ethics and the Environment* 22 (1): 63–87.

Mellor, Mary. 1992. *Breaking the Boundaries: Towards a Feminist Green Socialism*. London: Virago.

———. 1997. *Feminism and Ecology*. New York: New York University Press.

Merchant, Carolyn. 1992. *Radical Ecology: The Search for a Livable World*. New York: Routledge.

Mies, Maria, and Vandana Shiva. 1993. *Ecofeminism*. Halifax: Fernwood.

Mills, Bruce. 1994. *Cultural Reformations: Lydia Maria Child and the Literature of Reform*. Athens: University of Georgia Press.

Nord, Deborah Epstein. 1995. *Walking the Victorian Streets: Women, Representation, and the City*. Ithaca: Cornell University Press.

Olmsted, Frederick Law. 1870. *Public Parks and the Enlargement of Towns*. Cambridge, MA: American Social Science Association. *HathiTrust*.

———. 1981. *Papers*. Vol. 2. Baltimore: The Johns Hopkins University Press.

Pratt, Scott L. 2004. Rebuilding Babylon: The Pluralism of Lydia Maria Child. *Hypatia* 19 (2): 92–104. *Academic Search Premier*.

Review of *Hobomok*. 1824. *North American Review* 19: 262–263. *Making of America*.

Roberts, Heather. 2004. The Public Heart': Urban Life and the Politics of Sympathy in Lydia Maria Child's Letters from New York. *American Literature* 76 (4): 749–775.

Rosenthal, Debra J. 2002. Floral Counterdiscourse: Miscegenation, Ecofeminism, and Hybridity in Lydia Maria Child's *A Roman of the Republic*. *Women's Studies* 31: 221–245. *Academic Search Premier*.

Stout, Janis. 1976. *Sodoms in Eden: The City in American Fiction before 1860*. Westport: Greenwood.

Tingley, Stephanie A. 1997. 'Thumping Against the Glittering Wall of Limitations': Lydia Maria Child's "Letters from New York". In *In Her Own Voice: Nineteenth-Century American Women Essayists*, ed. Sherry Lee Linkon, 41–59. New York: Garland.

Vagionos, Alanna, and Damon Dahlen. 2017. 89 Badass Feminist Signs from the Women's March on Washington. *Huffington Post*, 21 January. https://www.huffingtonpost.com/entry/89-badass-feminist-signs-from-the-womens-march-on-washington_us_5883ea28e4b070d8cad310cd

Wach, Howard M. 2005. A Boston Vindication: Margaret Fuller and Caroline Dall Read Mary Wollstonecraft. *Massachusetts Historical Review* 7: 3–35.

Wollstonecraft, Mary. 1792. *A Vindication of the Rights of Woman; with Strictures on Political and Moral Subjects*. Boston: Thomas and Andrews. *Internet Archive*.

Women's March. Guiding Vision and Definition of Principles. https://www.womensmarch.com/s/WMW-Guiding-Vision-Definition-of-Principles-d5tb.pdf

CHAPTER 5

The Commons

GEORGE PERKINS MARSH AND LIBERAL ENVIRONMENTALISM

Contemporary environmental rhetoric too often raises the specter of planetary collapse, then calls on us to change our daily habits at home. The problem with this kind of argument is not that it engages in scare tactics. After all, total ecosocial breakdown is a real possibility. The problem is that we are offered such laughably inadequate responses. Hundreds of millions live without clean water, so we must think locally and install dual-flush toilets. The average global mean temperature may rise by as much as 6 degrees Celsius by the end of the century, so we must do our part by taking shorter showers. Sea level may rise a full meter, so we must love Mother Earth by weather-stripping our doors and windows. This kind of disconnection between incisive analysis of a genuine crisis and absurd calls for individual responses is a lamentably consistent thread in the history of American environmentalism. One of the first and most influential examples is *Man and Nature; or, Physical Geography as Modified by Human Action*, published in 1864 by the Vermont entrepreneur, polymath, and diplomat, George Perkins Marsh. Like so many environmentalists today, he was both deeply insightful about patterns of ecosocial damage *and* flatly unimaginative about how to respond.[1]

On the one hand, Marsh was one of the earliest writers in the United States to argue that human societies have degraded natural systems on a planetary scale. Most geographers before him had only asked how

© The Author(s) 2019
L. Newman, *The Literary Heritage of the Environmental Justice Movement*, Literatures, Cultures, and the Environment,
https://doi.org/10.1007/978-3-030-14572-9_5

environmental conditions have "influenced the social life and social progress of man." Marsh made this inquiry dialectical by exploring the ways that "man has reacted upon organized and inorganic nature, and thereby modified, if not determined, the material structure of his earthly home."[2] This was a meaningful step forward. Not only did he recognize that humans more than any other animal change their surroundings to suit their needs and desires, he also saw that the "unforeseen though natural consequences" of this necessary labor can be quite destructive.[3] Marsh could be fiery in his condemnation of ecological destruction: "Man has too long forgotten that the earth was given to him for usufruct alone, not for consumption, still less for profligate waste... Man is everywhere a disturbing agent. Wherever he plants his foot, the harmonies of nature are turned to discords."[4] More often though, he discussed human-caused environmental change in value-neutral language and suggested that the unforeseen consequences might just be positive: "The action of man upon the organic world tends to subvert the original balance of its species, and while it reduces the numbers of some of them, or even extirpates them altogether, it multiplies other forms of animal and vegetable life."[5] After all, Marsh was not a deep ecologist. He believed that human beings belong "to a higher order of existences than those born of [Nature's] womb and submissive to her dictates."[6] Moreover, humans "cannot rise to the full development of their higher properties, unless brute and unconscious nature be effectually combated, and, in a great degree, vanquished by human art."[7] Marsh enthusiastically endorsed Enlightenment-era narratives of social progress, and he hoped that we might "become the architects of our own abiding place."[8] In order for human well-being to improve, "improvidence, wastefulness, and wanton violence" would need to be replaced by "foresight and wisely guided persevering industry."[9] He sounds a note of confident hope in *Man and Nature*, calling on his reader "to become a co-worker with nature" and to join him in recognizing "the necessity of restoring [its] disturbed harmonies."[10] Marsh transplanted the best features of New England Transcendentalism, namely, its spiritual confidence and its reverence for nature, into the ground of geographical science. Along the way, he developed a deep understanding of human impacts on the global environment and the need to bring them under rational control. This is no small intellectual accomplishment.[11]

On the other hand, Marsh embodied the deep ideological contradictions that defined antebellum New England's ruling class. During the century between the American Revolution and the Civil War, the Northern bourgeoisie was pulled simultaneously in progressive and reactionary

directions by the historical trends with which it contended. Its most immediate tasks were to build industry, infrastructure, and a large force of reliably subordinate wage laborers. At the same time, the campaign to integrate
the national economy by overthrowing slavery demanded at least a rhetorical commitment to the egalitarian principles of republicanism. This central
contradiction drove the invention of American liberalism, the powerful set
of ideas that would authorize the bourgeoisie's dominance from then until
now. Liberalism exists to balance irreconcilable opposites, such as worker's
rights and corporate power, or markets and the common good. *Man and
Nature* valorizes a middle landscape, a liberal pastoral *topos*, in which the
citizens of a republic mix their labor with the soil on farms that they own
as private property. The organic products of their solitary labor are collective prosperity and liberty. In this rural utopia, the sublimely monotonous
primitive forest that once covered the earth has been replaced by an aes-
thetically pleasing landscape of fields, fallows, and woodlots that are worked
by self-reliant yeomen who steward their freeholds wisely.

At any moment, though, the beauty of this groomed landscape may
give way to ruins and weeds. As an avid student of the Roman Empire,
Marsh looked to its history for clues to the future of the American experiment. *Man and Nature* opens with an introductory parable about the fall
of Rome that foreshadows the logic of the book as a whole. "At the period
of its greatest expansion," Marsh writes, the Roman Empire "comprised
the regions of the earth most distinguished by a happy combination of
physical advantages," including mild climate, fertile soil, varied cultivars,
and "natural facilities for the transportation and distribution of exchange-
able commodities."[12] Importantly, this cornucopia was not natural: "The
spontaneous nature of Europe, of Western Asia, of Libya, neither fed nor
clothed the civilized inhabitants of those provinces. Every loaf was eaten
in the sweat of the brow." For Marsh, labor is the diagnostic marker of
civilization. And what made the Roman Empire's spectacular growth possible is that "toil was nowhere else rewarded by so generous wages."[13] In
other words, a fortuitous conjuncture of ecological and social conditions
engendered an historical triumph.

Marsh next draws an invidious comparison with the landscape of
nineteenth-century Italy, where he was serving as ambassador while writing *Man and Nature*: "More than one half " of the former Empire's
domain is now "either deserted by civilized man and surrendered to hope-
less desolation, or at least greatly reduced in both productiveness and
population."[14] To prove this point, Marsh delivers a classic litany of environmental devastation:

Vast forests have disappeared from mountain spurs and ridges; the vegetable earth accumulated beneath the trees by the decay of leaves and fallen trunks, the soil of the alpine pastures ... are washed away; meadows, once fertilized by irrigation, are waste and unproductive ... rivers famous in history and song have shrunk to humble brooklets ... the beds of the brooks have widened into broad expanses of pebbles and gravel ... the entrances of navigable streams are obstructed by sandbars, and harbors, once marts of an extensive commerce, are shoaled by the deposits of the rivers [which] have converted thousands of leagues of shallow sea and fertile lowland into unproductive and miasmatic morasses.[15]

As a result of this spiraling erosional disaster, large parts of the Old World have been reduced to "a desolation almost as complete as that of the moon" and the earth as a whole "is fast becoming an unfit home for its noblest inhabitant."[16] Marsh concludes that the "vengeance of nature for the violation of her harmonies" threatens to result in "the depravation, barbarism, and perhaps even the extinction of the species."[17] Crucially, the ultimate "Causes of this Decay" are social and political:

the primitive source ... of the acts and neglects which have blasted with sterility and physical decrepitude the noblest half of the empire of the Cæsars, is, first, the brutal and exhausting despotism which Rome herself exercised over her conquered kingdoms ... then, the host of temporal and spiritual tyrannies which she left as her dying curse to all her wide dominion, and which, in some form of violence or of fraud, still brood over almost every soil subdued by the Roman legions.[18]

Marsh goes on to list specific manifestations of this tyranny: unfair taxes, military conscription, and vast public works that exhaust social resources, and he complains that Rome "hampered industry and internal commerce by absurd restrictions and unwise regulations." How did tyranny cause environmental destruction? "Man cannot struggle at once against crushing oppression and the destructive forces of inorganic nature. When both are combined against him, he succumbs ... and the fields he has won from the primeval wood relapse into their original state of wild and luxuriant, but unprofitable forest growth."[19] In other words, the growth of an authoritarian social order in Rome produced wastefulness and ecosocial collapse, both by preventing peasants, workers, and slaves from thinking beyond short-term survival and by encouraging conspicuous consumption among the wealthy. This is a thoughtful piece of historical analysis, but

Marsh's application of his insight is not nearly so astute. Contemplating the destruction of the Old World, he concludes that any environmental progress there "must await great political and moral revolutions in the governments and peoples by whom those regions are now possessed."[20] His tone makes clear that he does not expect to see progress in his lifetime. In the New World, though, he sees a fleeting opportunity and he appeals to the reader: "Let us be wise in time, and profit by the errors of our older brethren!"[21] In the United States, which he believes to be free of the burden of history, it will be possible to create "a well-ordered and stable commonwealth" that acts upon nature with a "self-conscious and intelligent will." The key to success is "the diffusion of knowledge ... among the classes that, in earlier days, subdued and tilled ground in which they had no vested rights, but who, in our time, *own their woods, their pastures, and their ploughlands as a perpetual possession for them and theirs*, and have, therefore, a strong interest in the protection of their domain against deterioration."[22] Marsh imagines a "well-ordered" agricultural landscape as a patchwork of privately owned farms whose owners make rational decisions in order to ensure the sustainable growth of their investments.

Ironically, Marsh's dedication to the principle of individual self-interest shows most clearly in his critique of the "associate action" of the period's growing capitalist firms. He argues that "private corporations – whose rule of action is the interest of the association, not the conscience of the individual ... may become most dangerous enemies to rational liberty, to the moral interests of the commonwealth, to the purity of legislation and of judicial action, and to the sacredness of private rights."[23] In classically republican terms, growing corporations threaten to impose a new form of tyranny. Marsh rejects direct government policing, though, and he acknowledges that in the face of corporate irrationality, "the action of ... local or general legislatures" has "been found impotent to prevent ... destruction, or wasteful economy, of public forests." Thus the "only legal provisions from which anything is to be hoped, are such as shall make it a matter of private advantage to the landholder to spare the trees upon his grounds."[24] The leaders, then, of "a people of progress" should be wise men "endowed with an intelligent public spirit,"[25] and they should counterbalance corporate power by encouraging the wide diffusion of private property, so that the enlightened self-interest of individual land owners will lead them to maintain the ecological health of the land. In short, despite the clarity of its understanding of human impacts on the environment, *Man and Nature* concludes with a classically liberal theory of ecosocial change that centers on the power of rational self-interest. In the century and a half since 1864, the environmental crisis has deepened by orders of magnitude, but one thing has remained constant in environmental rhetoric: the liberal pastoral vision of modernity as a community of freeholders that has been deranged by the "rottenness of private corporations" and that can be redeemed by "the conscience of the individual."[26] That is to say, the dominant thread in environmental discourse today continues to focus on individual ethics and behaviors despite the now unavoidable truth that both the ecological problems we face and their causes are planetary in scale. We continue to concentrate on changing our neighbors' hearts and habits even though we know that global warming is an effect of globalization, which can only be meaningfully challenged by a cooperative global response.

JOHN CLARE, HENRY THOREAU, AND WALKING THE COMMONS

John Clare and Henry David Thoreau offer an important alternative to Marsh's liberal pastoral vision.[27] There is no evidence that Clare was acquainted with Thoreau's work, which achieved very limited circulation in England before the 1890s. And while Thoreau may have encountered a few of Clare's poems in the many anthologies and magazines that he read, there is no indication in his journals or elsewhere that he took note of them. However, Clare and Thoreau were kindred spirits.[28] Walking was a lifelong practice for both. It shaped their thought and structured their art. More particularly, both developed immersive ways of writing about walking the commons that showed what kinds of vernacular experience were endangered by privatization.[29]

Both Clare and Thoreau were literary outsiders whose writing was shaped by their experience as laborers in rural communities that were being rapidly transformed by the rise of capitalism. Clare earned his wages in the fields of Northamptonshire during the period of enclosure, while Thoreau supported himself, as he put it, "by surveying, carpentry, and day-labor of various kinds" in Concord as it was being integrated into greater Boston.[30] Their shared experience as members of the rural working class may help explain their distinctive ways of writing about nature. For instance, both Clare and Thoreau were sometimes impertinently rustic, perhaps because, as outsiders, they recognized the arbitrariness and artificiality of the conventions that governed polite society and polite literature.

Thoreau's first biographer records Marston Watson's recollection that Thoreau attended chapel at Harvard University in a green coat—"green, I suppose, because the rules required black."[31] And John Clare, during his short-lived ascendancy as the "Green Man" of the literary salons of London, wore a "bright, grass-colored coat" that made him look "like a patch of turnips amidst stubble and fallow."[32]

A shared penchant for performing extravagantly down-to-earth identities carried over into their poetry and prose as well. Both Clare and Thoreau used distinctive vernaculars to authenticate literary personae that were endemic to specific locales. For instance, in "Emmonsails Heath in Winter," Clare writes, "I love to see the old heath's withered brake/ Mingle its crimpled leaves with furze and ling."[33] The choice of common names for birds, as well as the use of dialect words like "oddling" for "solitary" and "bumbarrel" for "tit," registers Clare's traditional ecological knowledge of his neighborhood, which was grounded in daily familiarity, not in the itinerant natural historian's method of observation and classification. When Thoreau, in "A Winter's Walk," describes a frozen wetland, the dialect is different, but the effect is the same:

> Our feet glide swiftly over unfathomed depths, where in summer our line tempted the pout and perch, and where the stately pickerel lurked in the long corridors, formed by the bulrushes. The deep, impenetrable marsh, where the heron waded and bittern squatted, is made pervious to our swift shoes, as if a thousand railroads had been made into it. With one impulse we are carried to the cabin of the muskrat, that earliest settler, and see him dart away under the transparent ice, like a furred fish, to his hole in the bank; and we glide rapidly over meadows where lately "the mower whet his scythe," through beds of frozen cranberries mixed with meadow-grass. We skate near to where the blackbird, the pewee, and the kingbird hung their nests over the water, and the hornets builded from the maple in the swamp.[34]

Common names and homely diction set the tenor of the passage and establish Thoreau's *bona fides* as a local. At the same time, the quotation from John Milton's "L'Allegro" juxtaposes Thoreau's colloquial tone to the elaborate formality of one of the most canonical examples of the courtly pastoral tradition. In short, both Clare and Thoreau were acutely conscious of—and artfully cultivated—their identities as ordinary country people with homegrown knowledge of actual nature, as opposed to literary Nature. They were not deceiving themselves or us, for both were self-taught ecologists whose journals and phenological calendars record their comprehensive understanding of the biotic communities they inhabited and studied.

Clare and Thoreau achieved their extraordinary familiarity with their native places during lives of regular walking. Both began walking innocently enough in their youth but soon found that they had taken up a practice that was much more absorbing than a mere pastime. In his fragmentary autobiography, Clare fondly recollects his habit of roving: "I always wrote my poems in the fields and when I was out of work I used to go out of the village to particular spots which I was fond of from the beauty or secrecy of the scenes or some association and I often went half a day's journey from home on these excursions."[35] In addition to being a key element of his poetic practice, walking was also a form of self-treatment that Clare undertook to address the mental illness that eventually led to his confinement.[36] Likewise, in the essay "Walking," Thoreau remarks, "I think that I cannot preserve my health and spirits unless I spend four hours a day at least – and it is commonly more than that – sauntering through the woods and over the hills and fields absolutely free from all worldly engagements."[37] Thoreau was not speaking casually about preserving his health. Throughout his adult life, he experimented with folk remedies, ranging from dietary changes to exercise to travel, in the hope of controlling the tuberculosis that eventually killed him. Daily walking was an especially congenial regimen since it engaged both body and mind. Eventually, then, for both writers, walking evolved from a healthful routine into a central feature of their literary vocation. Indeed, their shared habit of using terms of art like roving, rambling, and sauntering demonstrates their self-consciousness about walking as a deliberate creative practice that took place at its own pace and according to its own rules.

After all, how we move through space shapes our habits of mind. Both Clare and Thoreau came to understand the world from the moving perspective of the footpath. Moreover, both wrote about nature in texts that are structured, not by the typical device of the prospect view, but as ambulatory excursions across horizontal landscapes. Some of these excursion narratives are continuous accounts of specific rambles, such as Clare's "Recollections after an Evening Walk" and Thoreau's "A Walk to Wachusett." Others are discontinuous texts in which the structuring device of an excursion disappears almost entirely behind a seemingly chaotic stream of sensory imagery and lyric statement. John Barrell remarks on this feature of many of Clare's poems and suggests that each one records "a complex moment of pleasure, produced by the simultaneous coming-together of a manifold of impressions in the speaker." Barrell further suggests that Clare deliberately composed poems in which the speaking self does not differentiate itself from its surroundings and that these poems "are the products of a self-conscious attempt to invent a language to represent a mode of consciousness that is, as he put it, distinctively 'local.'"[38] In addition to enacting the immersive experience of locals in a pointed contrast with that of outsiders, this method of composition allows a kind of rhetorical fluidity. A text written in this nonlinear style can range effortlessly across descriptive, meditative, declamatory, and even hortatory modes. That is to say, many of Clare's poems and many of Thoreau's writings enact the experience not just of wandering, but of sauntering *thoughtfully*. Experiences give rise to ideas which in turn provoke declarations.

Clare's "Helpstone," for instance, is centered in a present moment that is first revealed some 50 lines into the poem, when the speaker exclaims:

> Hail scenes obscure so near and dear to me
> The church the brook the cottage and the tree
> Still shall obscurity rehearse the song
> And hum your beauties as I stroll along[39]

At the poem's opening, this generative occasion is invisible; instead, the first lines enter directly into the current of thought that occurred during the walk that provoked the poem's composition. As a result, Clare is able to leap effortlessly from sympathizing with "little birds in winters frost and snow" to meditating on the universality of "fondness" for one's "native place" to describing beetles dancing on the surface of a stream to lamenting "the woodman's cruel axe" to marveling at a pasture bedecked with kingcups and cowslaps to denouncing "accursed wealth" for starving the poor and leveling the woods. Many of Clare's poems, such as "Evening," "The Gipseys Camp," "Recollections after a Ramble," "Emmonsales Heath," and "Remembrances," rove in similar ways.

Likewise, in many of Thoreau's excursions, such as *A Week on the Concord and Merrimack Rivers* (1849) and *Cape Cod* (1865), the narrative serves as little more than the occasion for a stream of sensation and cognition that it does not overdetermine. These texts are placed, but they are not superintended by—nor do they superintend—the places where they take place. Instead, they embody the experience of moving across the landscape at ground level, perceiving it close up, and steering by impulse and interest. In "Walking," Thoreau describes this kind of "sauntering" as an art for which only a few people have "a genius."[40] Walking is a way to "shake off the village," to forget "church and state and school, trade and commerce, and manufactures and agriculture, even politics," in order to reconnect with the "Wildness [that] is the preservation of the world."[41] Characteristically, he applies this doctrine of wildness to writing and makes a succinct statement of organic literary aesthetics: "A truly good book is something as natural, as unexpectedly and unaccountably fair and perfect, as a wild flower discovered on the prairies of the West or in the jungles of the East."[42] Because he consistently hove to this literary creed, even his most village-bound essays, like "Civil Disobedience" and "Life Without Principle," ramble improvisationally through the fields of nineteenth-century political and philosophical discourse.

Walking traditional footpaths that traverse property lines—Thoreau calls these "across-lot routes"—tends to make one aware not only that legal boundaries fail to correspond to the contours of the ecosocial communities on which they have been imposed but also that those boundaries are arbitrary and even capricious.[43] This is an empowering revelation, since, after all, the central claim of private property as a social form is that it was ever thus.[44] Both Clare and Thoreau opposed privatization of public lands; however, for different reasons, they rarely wrote about the issue. Clare and his family were dependent on the patronage of the very agricultural capitalists who were driving the process of enclosure. One of them, Lord Radstock, famously scribbled the phrase, "This is radical slang," in the margin of an early draft of *The Village Minstrel* and relentlessly pressured Clare to censor himself.[45] Thoreau likely censored himself for a time because he believed that it "is not a man's duty, as a matter of course, to devote himself to the eradication of any, even the most enormous wrong."[46] When he later abandoned this view, the enormous wrong he dedicated himself to eradicating was slavery. Nevertheless, both Clare and Thoreau, as thoughtful walkers, did speak out in key texts against enclosure of the commons.[47]

In Clare's great elegy, "The Mores," a personified "travel," rambles through the countryside where post-enclosure fields in which "men and flocks [are] imprisoned ill at ease" contrast with the "unbounded freedom" of what were formerly commons. The poem begins with a distant prospect of a nature that is pristine across both space and time:

> Far spread the moorey ground a level scene
> Bespread with rush & one eternal green
> That never felt the rage of blundering plough
> Though centurys wreathed springs blossoms on its brow[48]

In this long view, the pre-enclosure moors seem to be a conventionally awe-inspiring scene—boundless, uniform, and grand. But in the next few lines, it becomes clear that Clare is not just recycling the usual idioms of sublimity:

> Still meeting plains that stretched them far away
> In uncheckt shadows of green brown & grey
> Unbounded freedom ruled the wandering scene
> Nor fence of ownership crept in between
> To hide the prospect of the following eye[49]

Clare reverses the historical narrative that justified enclosure. In that tale, wild and sublime landscapes were tamed and beautified by conversion to for-profit agricultural production. But Clare asserts that the moors were the historical site of a freedom that was destroyed when the land was divided into privately owned and geometrically fenced tracts. Before enclosure, the land's "only bondage was the circling sky."[50] Moreover, Clare makes clear that he is not concerned merely with symbolic freedoms but with concrete matters of access to the land:

> Now this sweet vision of my boyish hours
> Free as spring clouds & wild as summer flowers
> Is faded all – a hope that blossomed free
> & hath been once no more shall ever be
> Inclosure came & trampled on the grave
> Of labours rights & made the poor a slave[51]

Eighteenth-century agricultural reformers justified enclosure by describing it as a process of bringing waste lands and barren fields into production. But traditional use rights on the commons were already producing crucial food and fuel for the rural poor. Enclosure not only changed the face of the land, it also extinguished their rights of access, forcing more and more people to survive by working for wages. For many, this loss of freedom on the land felt like a descent into bondage. Clare offers a retrospective vision of the kind of liberty that has been taken away:

> The sheep & cows were free to range as then
> Where change might prompt nor felt the bonds of men
> Cows went and came with evening morn and night
> To the wild pasture as their common right
> And sheep unfolded with the rising sun
> Heard the swains shout and felt their freedom won.[52]

What is at issue here is, of course, the tradition of ranging stock collectively, combining the few animals owned by each of a village's families into a single large flock that was managed collectively. But Clare does not take time to compare the efficiency of feudal and capitalist agricultural practices. Instead, he ignores the strictly utilitarian criteria of judgment established by the improvers, and he denounces his loss of access to the materials of aesthetic experience. In other words, he protests the loss of what, in "The Progress of Rhyme," he calls the "right to song."[53] Clare's art depended on his vernacular knowledge of the land, which depended in turn on the freedom to travel into the remote corners of his "native plain," which, he believed, should be "a garden free for all."[54] Clare rises, in the final lines of "The Mores," to a powerful condemnation of privatization:

> These paths are stopt – the rude philistines thrall
> Is laid upon them & destroyed them all
> Each little tyrant with his little sign
> Shows where man claims earth glows no more divine[55]

The epithet "rude philistines" is decisive. "The Moors" demonstrates that, in Raymond Williams's words, "Wealth is not only hard and cruel but tasteless."[56] The structures of land tenure and the relations of agricultural production imposed by the improvers have directly cut the rural poor off from the land where, yes, they supplemented the uncertain fruits of their labor, but more importantly, where, in doing so, they demonstrated their freedom and independence of mind by appreciating the beauty of commons that only they knew so well.

Among the papers that Thoreau left at his death in 1862 was the manuscript of *Wild Fruits*, a field guide to the plants of New England that produce edible berries, nuts, and seeds. There are more than 100 entries that describe plants ranging from strawberries, choke cherries, and sassafras to black oak, beech, and walnut. Many of the entries spill over into excursive commentary on a range of topics, including plant succession after forest fires, Native American use of wild berries, the respective dietary habits of Old and New England, the relative value of common and scientific names, and the time a farmer paid him in red huckleberries for surveying a field. Thoreau uses his entry on "Black Huckleberry" as an occasion to address the privatization of the commons:

> What sort of a country is that where huckleberry fields are private property? When I pass such fields on the highway, my heart sinks within me. I see a blight on the land. Nature is under a veil there. I make haste away from the

desecrated spot. Nothing could deform her fair face more. I cannot think of it ever after but as the place where fair and palatable berries are converted into money, where the huckleberry is desecrated. It is true, we have as good a right to make berries private property as to make wild grass and trees such.[57]

The wild huckleberry stands here as a synecdoche for open country, and Thoreau goes on to make clear that, like Clare, his concerns are not limited to traditional gathering rights. The problem with fencing off the land into tracts of private property is that when "we exclude mankind from gathering berries in our field, we exclude them from gathering health and happiness and inspiration [...]."[58] At a time when the common school movement was successfully establishing the value of community investment in education, Thoreau, a former teacher, draws an analogy between a town's open land and its public schools:

I well remember with what a sense of freedom and adventure I used to take my way across the fields with my pail [...] toward some distant hill or swamp, when dismissed for the day, and I would not now exchange such an expansion of all my being for all the learning in the world. Liberation and enlargement – such is the fruit which all culture aims to secure. I suddenly knew more about my books than if I had never ceased studying them. I found myself in a schoolroom where I could not fail to see and hear things worth seeing and hearing, where I could not help getting my lesson, for my lesson came to me. Such experience, often repeated, was the chief encouragement to go to the Academy and study a book at last.[59]

Given his belief that common lands provided invaluable bodily, intellectual, and even spiritual sustenance, Thoreau called for their public management. At the end of the *Wild Fruits* manuscript, there is an account of an excursion to a nearby town, which appears to be a draft of the conclusion for the book and which may also have been meant for the meat of a lyceum lecture. "I do not believe," Thoreau begins, "that there is a town in this country which realizes in what its true wealth consists. I visited the town of Boxboro only eight miles west of us last fall, and far the handsomest and most memorable thing which I saw there was its noble oak wood."[60] Thoreau ridicules an imaginary town historian for finding this oak wood less interesting than "the history of the parish," and he laments the fact that it will "be cut off within a few years for ship-timber and the like."[61] Finally, he states, somewhat drily, "It is too precious to be thus disposed of."[62] Thoreau rises from this particular example to a general statement of principle:

What are the natural features which make a township handsome and worth going far to dwell in? A river with its waterfalls, meadows, lakes, hills, cliffs, or individual rocks, a forest and single ancient trees. Such things are beautiful. They have a high use which dollars and cents never represent. If the inhabitants of a town were wise, they would seek to preserve these things, though at a considerable expense. For such things educate far more than any hired teachers or preachers, or any at present recognized system of school education.[63]

In the pages that follow, Thoreau argues for public ownership and management of not only forests but also prominent hilltops and mountains, as well as rivers and streams, including their banks. These places should be held as "common possession[s] forever, for instruction and recreation."[64] He brings his argument to a ringing conclusion by returning to his original analogy: "We boast of our system of education, but why stop at schoolmasters and schoolhouses? We are all schoolmasters, and our schoolhouse is the universe."[65]

In the end, what sets Clare and Thoreau apart from their contemporaries is that they write as though their minds were properties of physical bodies walking in their native places. Instead of ascending mountains in solitude to survey the homogenized and dominated landscapes of the sublime and the picturesque, they move horizontally through complex, patchy, and locally specific landscapes. Their densely imagistic poetry and prose move beyond Romantic idealizations of pristine wilderness to capture the detailed materiality of their home places, working landscapes where human beings are part of a combined and indissoluble ecosocial community. Moreover, intimate knowledge of the local community grounds their identification—they would have called it sympathy—with other beings, whether human, animal, or vegetable, which in turn inspires their ethical commitment to the community's well-being. Clare and Thoreau anticipated by several decades the core environmentalist arguments that there are higher uses of the land than to exploit its fertility for profit and that access to wild nature is a fundamental human right. Just as importantly, these kindred spirits experimented with a variant of the radical pastoral *topos* that was rooted in local knowledge and that inspired them to protest the exploitation of the commons for profit. Clare and Thoreau are increasingly relevant today, at a time when privatization continues to fragment our ecosocial communities. Now more than ever, we need the commons as a place of learning and inspiration. Walking the land together can help us see our world more clearly, commit to change it, and understand how.

THOREAU'S MATERIALISM AND ENVIRONMENTAL POSSIBILISM

Walking changed Thoreau from an idealist to a materialist whose ways of thinking about ecosocial change offer an important alternative to the liberal environmental tradition. In an 1842 lecture, "The Transcendentalist," Ralph Waldo Emerson contrasted materialism with the idealism that he and his circle espoused.[66] "The materialist," he says, "insists on facts, on history, on the force of circumstances, and the animal wants of man; the idealist on the power of Thought and of Will, on inspiration, on miracle, on individual culture."[67] In other words, a Transcendentalist was an idealist who particularly believed in the self-reliant human spirit's absolute independence from the body and nature. For almost a century after Thoreau's death, most readers took it for granted that, like his mentor, he was an idealist. If he often wrote in vivid sensory detail about nature, it was because he sought to restate Transcendentalist ideas in emblematic particulars. As Emerson put it shortly after Thoreau's death: "In reading him, I find the same thought, the same spirit that is in me, but he takes a step beyond, & illustrates by excellent images that which I should have conveyed in a sleepy generality."[68] Alongside this dominant perception of Thoreau as a poetic idealist with a talent for finding correspondences between natural facts and spiritual truths, there has always been a second perspective. Most of those who knew him personally (rather than through his writings and reputation) saw him mainly as a literary natural historian with a remarkable talent for close observation and engaging description.[69] The reality is that Thoreau's thinking evolved over time. In particular, his understanding of the human community evolved in parallel with his insight into nonhuman nature, which formed during his across-lot walks. He increasingly applied a natural historian's habits of mind—empirical observation and materialist analysis—to the social and political life of Concord, and by extension, to the antebellum United States as a whole. As a result, he saw that capitalism actively managed the relationship between humans and nature by organizing labor according to the economic forms of property and profit. The exploitation of labor (both wage and slave) and the appropriation of nature were thus twin features of an intensely destructive modernity. He also acknowledged the need not just for individual self-reform but also for collective action in pursuit of wholesale social change. In other words, Thoreau was a materialist, not just in the philosophical sense but also in the political sense. He believed that not just our ideas but the socioeconomic relationships that shape our lives on the land must be changed if we are to achieve a just society in a healthy environment.[70]

In *Walden*, Thoreau redeveloped the ancient idea that higher human functions are impossible if basic needs have not been fulfilled, and in doing so, he suggested that bodily and mental being are tightly interwoven. At the same time, Thoreau repurposed the ideas of Arnold Guyot and other mid-nineteenth-century geographers, who applied materialist and empiricist ways of thinking to questions about the relations between humans and their environments. In contrast with their environmental determinism, which functioned as a racist apology for colonialism, Thoreau offered a *possibilistic* way of thinking about the relationships between societies and their geographical settings; that is, he emphasized the human capacity to adapt creatively to varying material conditions. *Walden* demonstrates that the processes by which we fulfill our basic bodily needs, as well as the lives of the mind and spirit that our bodies make possible, are both constrained and activated by the ecosocial environments that we create and within which we live. As Emerson put it, Thoreau "insists on facts, on history, on the force of circumstances, and the animal wants of man."[71]

The opening pages of *Walden* establish what appears to be a very clear ranked opposition between the material and the ideal. Thoreau directly states that the book's topic is our "*outward* condition or circumstances in this world."[72] More specifically, he announces that he will address the problem of incessant labor to service debt, and then he uses a jarring phrase, "serfs of the soil," to describe his proverbially self-reliant New England neighbors. This aggressive opening gambit is followed by a series of grotesque images of close bodily intimacy with the land. Farmers "eat their sixty acres" and "begin digging their graves as soon as they are born."[73] So far, "this world" seems sordid enough. But soon, a scatological joke identifies the problem even more specifically: "The better part of man is soon ploughed into the soil for compost." That is, excessive labor turns consciousness into an excremental by-product. The human faculties of thought and will are degraded when they used for no more noble purpose than growing commodities for profit. Thoreau's tone turns from sarcastic to serious as he develops this point: "The laboring man has not leisure for a true integrity day by day.... He has no time to be anything but a machine."[74] Not only do the "superfluously coarse labors of life" debase our intellect, degrade our moral courage, and compromise our self-determination, they also prevent us from enjoying life's "finer fruits," such as spiritual and aesthetic experiences.

Thoreau is not condemning labor *per se*. Rather, he makes clear that the "finest qualities of our nature" are destroyed by meaningless work that we undertake "to get out of debt, a very ancient slough."[75] The allusion here to John Bunyan's *Pilgrim's Progress* (1678), in which Christian sinks into the Slough of Despond under the weight of his own sins, suggests a parallel between debt and immobilizing guilt, and it reinforces the idea that financial entanglements prevent us from entering the Celestial City. Thoreau declares in frustration: "Talk of a divinity in man! Look at the teamster on the highway, wending to market by day or night; does any divinity stir within him? His highest duty [is] to fodder and water his horses! What is his destiny to him compared with the shipping interests? Does not he drive for Squire Make-a-stir? How godlike, how immortal, is he?"[76] This passage draws a bright line between the material world of debt, work, and soil and the ideal world of eternal spiritual life.

However, as soon as Thoreau finishes walling off this border, he begins to dance across it. He announces that he will experiment with alternatives to "the common mode of life" and then introduces a new metaphor: "The incessant anxiety and strain of some is a well nigh incurable form of disease."[77] Thoreau knew all too well what it meant to live with an incurable bodily disease that could slow his mind and depress his spirits. His daily regimen of walking and writing, which he pursued under the ever-present threat of death, was designed to maintain the vitality of his whole self. And his flat tone in this sentence suggests that he is no longer speaking figuratively and humorously; instead, he is acknowledging that bodily and mental health are not distinct phenomena, but braided processes. As it turns out, the idea that there is a fluid connectivity between the material and ideal worlds permeates "Economy." Thoreau begins to expand this thought by restating an old idea: "The necessaries of life for man in this climate may, accurately enough, be distributed under the several heads of Food, Shelter, Clothing, and Fuel; for not till we have secured these are we prepared to entertain the true problems of life with freedom and a prospect of success."[78] Three of these basic needs—clothing, shelter, and food—form the subjects of major subsections of "Economy," in which Thoreau shows that the way we fulfill them is embedded in a complex social and historical context. In each section, Thoreau ridicules absurd conventions and traditions and then explains the economic structures and social processes, such as financial debt and conspicuous consumption, that drive people to behave so irrationally. For instance, the first of these sections discusses clothing and begins with a bit of deadpan sarcasm: "perhaps we are led oftener by the love of novelty, and a regard for the opinions of men, in procuring it, than by a true utility."[79] After reminding us that "the object of clothing is, first, to retain the vital heat [of the body], and secondly, in this state of society, to cover nakedness," Thoreau mocks people who wear clothes as fashionable ornaments; explains how clothing is too often used to exhibit the wearer's wealth and social power; defends the practicality and honesty of cheap, sturdy garments; derides people who sneer at patches and old clothes; and finally, denounces the "factory system" that exploits underpaid workers to mass produce trivialized commodities.[80] Similarly, the section of "Economy" that discusses shelter contrasts the simple physical uses of houses with their symbolic functions in a hierarchical society, and then it suggests a materialist explanation for the misalignment. The basic function of a home, Thoreau writes, is to serve as "a place of warmth, or comfort, first of physical warmth, then the warmth of the affections."[81] However, many people waste their lives working to pay for absurdly large and luxurious houses both because they also serve as markers of status and because investment in real estate is a means of accumulating capital. Meanwhile, since wealth and property are distributed so unequally, workers and slaves must endure housing so poor that "the development of all their limbs and faculties is checked."[82] Human beings subsist, that is, in class-specific environments that can either constrain or support their bodily health and mental well-being.

Thoreau's famous response to the problems that he identifies in "Economy" comes in the form of a clear imperative: "Simplicity, simplicity, simplicity!"[83] However, when he recommends a simple life, he does so not because bodily needs are low or trivial, but because the way we fulfill them is deranged: "The farmer is endeavoring to solve the problem of a livelihood by a formula more complicated than the problem itself. To get his shoestrings he speculates in herds of cattle."[84] An irrational socioeconomic order encourages us to waste our best energies pursuing perverse aims, it implicates us in grave economic injustices, and it compels us to labor by ensnaring us in debt. Thoreau insists, though, that we have the capacity to "give up our prejudices."[85] After all, rather than see basic needs as an unchanging biological substrate of mental life, he understands them as material imperatives to which people have responded variably over the course of human evolution: "Man has invented, not only houses, but clothes and cooked food; and possibly from the accidental discovery of the warmth of fire, and the consequent use of it, at first a luxury, arose the present necessity to sit by it. We observe cats and dogs acquiring the same second nature."[86] The ways that we generate and maintain our "animal heat" vary according to natural environment and climate: "In cold, weather we eat more, in warm less." Our mode of living also varies according to social environment: "The poor man is wont to complain that this is a cold world and to cold, *no less physical than social*, we refer directly a great part of our ails."[87] Thus, when Thoreau lists the necessaries of life, he explicitly specifies that they are necessary "in this climate."[88] Similarly, he states that the goal of his experiment is to determine how best to live "at the present day, and in this country."[89]

In "Economy," Thoreau is working both with and against the ideas of Arnold Guyot, the Swiss geologist and geographer who was brought to Massachusetts by the Harvard zoologist, Louis Agassiz.[90] Guyot's 1849 book, *The Earth and Man: Lectures on Contemporary Physical Geography in its Relation to the History of Mankind*, was one of the most influential scientific works of its century, in large part for the very reasons that it is now forgotten. Guyot argued that Europe and North America were "inhabited by the finest races" because human beings had adapted there to a temperate climate where it took hard work and planning to fulfill basic needs.[91] At the same time, according to Guyot, the overly cold, dry, or otherwise exhausting conditions that prevailed in most other regions of the world had caused the degeneration of the "races of man" that lived there. Thus, the "man of the polar regions is [a] beggar, overwhelmed with suffering, who, too happy if he but gain his daily bread, has no leisure to think of anything more exalted."[92] Meanwhile, the overly fertile and productive climates of the equatorial zone made it possible to survive there without the effort or forethought that spurred intellectual growth: "Thus, if the tropical continents have the wealth of nature, the temperate continents are the most perfectly organized for the development of man. They are opposed to each other, as the body and the soul, as the inferior races and the superior races, as savage man and civilized man, as nature and history."[93] Guyot was not just casually or incidentally racist. He explicitly rejected Rousseauian primitivism, with its image of the noble savage: "the precious treasures of intelligence and of lofty thoughts contained in our libraries, – where would they be if human societies had retained that simplicity which a false philosophy has called the simplicity of nature [...]?" Moreover, he measured a society's level of civilization according to the diversification of its forms of labor and the variety of its commodities. Thus, he praised the "thousand wants of a society as complicated as ours" since they were "the sign of a social state arrived at a high degree of improvement."[94] *The Earth and Man* conveniently justified the technological, economic, and military dominance of Europe and the United States during the colonial period in terms of an environmental-determinist theory of racial difference.

Thoreau read *The Earth and Man* in January and February 1851, during the two-year hiatus between his third and fourth revisions of *Walden*.[95] The two books provide alternative answers to the same question, which Thoreau frames this way: "What is the nature of the luxury which enervates and destroys nations?"[96] *Walden* returns repeatedly to the distinction between primitive and civilized states of social being. At times, it even seems to reproduce Guyot's racist categories uncritically, such as when Thoreau remarks on the purported insensitivity to extreme temperatures of the aboriginal inhabitants of Australia and Tierra del Fuego and then wonders whether it would be "impossible to combine the hardiness of these savages with the intellectualness of the civilized man?"[97] However, it soon becomes clear that Thoreau is, in fact, reversing Guyot's racial hierarchies. For instance, during his discussion of clothing, Thoreau writes,

> The very simplicity and nakedness of man's life in the primitive ages imply this advantage at least, that they left him still but a sojourner in nature. When he was refreshed with food and sleep he contemplated his journey again. He dwelt, as it were, in a tent in this world, and was either threading the valleys, or crossing the plains, or climbing the mountain tops. But lo! men have become the tools of their tools. The man who independently plucked the fruits when he was hungry is become a farmer; and he who stood under a tree for shelter, a housekeeper.[98]

Whereas Guyot maintained that primitive people were mired in an all-consuming struggle for survival and lived a merely bodily existence while civilized people enjoyed leisure which made possible a cultivated life of the mind and spirit, Thoreau suggests that the opposite is true: primitivism is an attribute of historical periods, not peoples. Further, in the Euro-American civilization that Guyot regarded as the most advanced on earth, basic needs must be satisfied with commodities produced for profit, and therefore *"the civilized man [...] is employed the greater part of his life in obtaining gross necessaries and comforts merely."*[99] To make matters worse, this never-ending work to fulfill basic needs too often fails to achieve its end. The very people "by whose labor the works which distinguish this generation are accomplished" must live "in sties, and all winter with an open door, for the sake of light, without any visible, often imaginable, wood pile, and the forms of both old and young are permanently

contracted by the long habit of shrinking from cold and misery [...]."
According to Thoreau, European and American workers and slaves suffer
lives that are materially worse than those of "the North American Indian,
or the South Sea Islander, or any other savage race before it was degraded
by contact with the civilized man."[100]

Thoreau's point in turning Guyot's ideas upside down is not to rein-
state primitivism. Instead, by documenting the equatorial life that Thoreau
lives in Concord, *Walden* demonstrates that one of Guyot's main premises
is false: a temperate climate does not necessarily "incite man to a constant
struggle."[101] After claiming counterfactually that he speaks "as one not
interested in the success or failure of the present economical and social
arrangements," Thoreau proves by experiment that "that it would cost
incredibly little trouble to obtain one's necessary food, *even in this lati-
tude*."[102] He finds that he can "do all his necessary farm work as it were
with his left hand at odd hours in the summer."[103] More than simply show
that there is nothing geographically inevitable about the common mode
of life in New England, Thoreau maintains that human societies can evolve
in response to changing socio-environmental conditions: "Nature and
human life are as various as our several constitutions" because "Man is an
animal who more than any other can adapt himself to all climates and circumstances."[104] That is to say, Thoreau envisions a wide range of both
actual and possible adaptive strategies across the span of human history. In
the abstract, no strategy is superior to any other; it can be evaluated only
in relation to the specific socio-environmental conditions within which it
operates as a means to fulfill human needs sustainably. One sure mark of a
society's adaptive failure is that, like the antebellum United States, it produces material inequality and poverty. In "the savage state every family
owns a shelter as good as the best, [but] in modern civilized society not
more than one half the families own a shelter" at all.[105] Thoreau is not
merely making an invidious comparison. Poverty and excessive labor prevent many people in so-called civilized society from fulfilling their human
potential, and as a result, "we are still forced to cut our *spiritual* bread far
thinner than our forefathers did their wheaten."[106] In a society that boasts
of its technological advances and material productivity, Thoreau asks,
"Why has man rooted himself thus firmly in the earth, but that he may rise
in the same proportion into the heavens above?"[107] Rather than present
the material and the ideal as opposite and incommensurate, Thoreau uses
this metaphor of arboreal growth to suggest that in a healthy society which
had adapted successfully to its environment, the material and the ideal
would be integrated into a single organic life process.

Walden grows according to just this arboreal pattern: "Economy" roots the book and Thoreau's individual life of the mind firmly in the earth, so that the chapters that follow may develop an integrated account of the ecosocial world for which Concord stands. Thoreau's experiment has allowed him to fulfill the "necessaries of life" without excessive labor, which in turn has made it possible for him "to front only the essential facts of life."[108] Among those essential facts, one of the foremost is that industrial capitalism, which he discusses via the synecdoche of the railroad, was driving a process of combined socio-environmental damage. As Leo Marx observed, the railroad is a constant presence in *Walden*, relentlessly disrupting Thoreau's pastoral retreat.[109] This should come as no surprise, though. Thoreau was a product of the years after the War of 1812 when the United States began its rise to world economic dominance. The second Bank of the United States was founded just before his birth; excavation of the Erie Canal began the week he was born; and during his childhood the United States saw the swift growth of factory production of textiles, farm implements, firearms, steam engines, and much more. Thoreau was, in other words, a witness to the birth of our world. And according to *Walden,* the economy, whose engine was the railroad, "that last improvement in civilization," had already become a fetish in its own right.[110] People had developed "an indistinct notion that if they keep up this activity of joint stocks and spades long enough all will at length ride somewhere, in next to no time, and for nothing."[111]

Thoreau lampoons blind faith in capitalism in the chapter "Sounds," where the railroad becomes the centerpiece of a spectacle of meaningless commerce:

The whistle of the locomotive penetrates my woods summer and winter, sounding like the scream of a hawk sailing over some farmer's yard, informing me that many restless city merchants are arriving within the circle of the town. [...] As they come under one horizon, they shout their warning to get off the track to the other, heard sometimes through the circles of two towns. Here come your groceries, country; your rations, countrymen! [...] And here's your pay for them! screams the countryman's whistle; timber like long battering rams going twenty miles an hour against the city's walls, and chairs enough to seat all the weary and heavy laden that dwell within them.[112]

Not only is this parade feverish and noisy, it is environmentally destructive: "All the Indian huckleberry hills are stripped, all the cranberry meadows are raked into the city."[113] Moreover, "when the smoke is blown away and the vapor condensed, it will be perceived that a few are riding, but the rest are run over."[114] Many of those whom the railroad runs over are the very people who lay the track: "Did you ever think what those sleepers are that underlie the railroad? Each one is a man."[115] However, the railroad's human cost is not limited to the exploitation of workers. A society's priorities have been distorted when it expends its best energies and ingenuity on mere commodities: "Up comes the cotton, down goes the woven cloth; up comes the silk, down goes the woollen; up come the books, but down goes the wit that writes them."[116] The economic activity that is necessary to fulfill basic material needs has eclipsed all other concerns and become an end in its own right, so that now we "do not ride on the railroad; it rides upon us."[117] The capitalist economy has become a powerful material force that directs our lives and may determine our destiny: "We have constructed a fate, an *Atropos*, that never turns aside."[118]

The materialist Thoreau, who both celebrated the body's sensory immersion in nature and recognized the materiality of the capitalist economy, should remain an important literary touchstone for environmental humanists today as we work toward an integrative understanding of the connections between the destruction of nature, the exploitation of labor, and racial and sexual oppression. We should also find inspiration in the creativity of his efforts to change the world he understood so clearly. When he awoke from the liberal dream of Transcendentalism into the material light of morning at Walden Pond, he saw clearly that his life of the embodied mind in nature occurred within a socio-environmental context that was fundamentally unjust. The pressing issue of his day, of course, was slavery. He saw that slavery was absolutely integral to American capitalism, not simply a residuum of feudalism or a peripheral quirk. He also saw that it amounted to both a violation of human rights on a colossal scale and an environmental catastrophe. As James Finley puts it, Thoreau came to understand that slavery "pollutes everything, including agricultural land, wilderness, labor conditions, politics, and interpersonal relationships."[119] More than simply deplore injustice, he *acted* to redress it—not only in his individual body and home but also in speech and writing, in acts of communication that were designed to convene new communities of resistance and even revolution. At a time when others were wearing linen instead of Southern cotton and making beet sugar to substitute for the cane sugar produced by slaves, he experimented with growing his own food and eating a vegetarian diet as a way of detaching himself bodily from the antebellum nexus of oppression and exploitation for profit.[120] He also engaged in more direct action. For instance, while he makes scant mention of slavery in *Walden,* he had reason to be discrete: his family home was the Concord stop on the Underground Railroad, and he frequently helped fugitive slaves make their way north to freedom.[121] Finally, during his daily walks across lots, he often encountered physical reminders of the history of free and enslaved people of color on the land in and around Concord, so he took pains to reconstruct and tell their stories in *Walden.*[122] Of course, he also spoke publicly against slavery whenever the opportunity presented itself, and he published multiple abolitionist essays based on his speeches. While he was campaigning for social justice, Thoreau also began to amass the data and develop the arguments for the preservation of public green space and biological reserves that he doubtless would have pressed had he lived beyond emancipation and the end of the Civil War. In a word, Thoreau's materialism, in both his life and works, reminds us that a vibrant life of the mind and spirit requires bodily well-being which in turn requires a just social order on a healthy planet. Moreover, our struggle to preserve, protect, and restore the global commons is a struggle for freedom and equality.

NOTES

1. For general introductions to Marsh and his work, see Dorman and Lowenthal.
2. Marsh, 13.
3. Marsh, 19.
4. Marsh, 36.
5. Marsh, 3.
6. Marsh, 37.
7. Marsh, 38.
8. Marsh, 14.
9. Marsh, 19.
10. Marsh, 35 and 12.
11. Among his many other progressive proposals, Marsh made an early and influential call for the designation of national parks on federal and state land: "It is desirable that some large and easily accessible region of American soil should remain, as far as possible, in its primitive condition, at once a museum for the instruction of the student, a garden for the recreation of the lover of nature, and an asylum where indigenous tree, and humble plant that loves the shade, and fish and fowl and four-footed beast, may dwell and perpetuate their kind, in the enjoyment of such imperfect protection as the laws of a people jealous of restraint can afford

them" (203). Of course, Marsh was an enthusiastic manifest destinarian, so there was no room for indigenous people in the parks he imagined, only indigenous trees. His vision of parks as carefully curated spaces where nonhuman beings are protected from human violence relied on a symbolic erasure of Native Americans that would all too often materialize in the years to come. America's best idea required not only the idea of America but also its violent creation.

12. Marsh, 7.
13. Marsh, 8.
14. Marsh, 9.
15. Marsh, 9.
16. Marsh, 42.
17. Marsh, 188 and 43.
18. Marsh, 11.
19. Marsh, 12.
20. Marsh, 45–46.
21. Marsh, 198.
22. Marsh, 46, my emphasis.
23. Marsh, 51–52n.
24. Marsh, 259 and 202.
25. Marsh, 280 and 202.
26. Marsh, 51n.
27. This section was previously published as Newman, "John Clare, Henry David Thoreau, and Walking."
28. I have borrowed this phrase from Peck, who discusses the formal and ideological similarities between Thoreau's writings and the canvases of Hudson River School painter, Asher Durand.
29. In the first article to explore the parallels between Thoreau and Clare, Markus Poetzsch focuses on their writings about birds and argues that they "not only wrote about nature from a related set of convictions but [also] perceived and moved through their respective environments, their topographies of home, with a kindred attunement to the ... otherness of nonhuman life" (92). Only a few others have connected Clare and Thoreau. Rayment's unpublished doctoral dissertation, "Empiricism and the Nature Tradition," compares both writers to Izaak Walton. In *Romantic Ecology*, Bate refers several times to Clare and briefly mentions Thoreau, but only implies a comparison. McKusick devotes individual chapters to Clare and Thoreau while tracing the literary history of modern ecological understanding from the British Romantic poets to the American Transcendentalists. And Pawelski and Moores offer consecutive chapters on Thoreau and Clare in a book that discusses the role of literature in the pursuit of the good life.

30. Thoreau, *Walden*, 58. When asked in 1847 to identify his occupation, Thoreau responded, "I am a Schoolmaster – a Private Tutor, a Surveyor – a Gardener, a Farmer – a Painter, I mean a House Painter, a Carpenter, a Mason, a Day-Laborer, a Pencil-Maker, a Glass-paper Maker, a Writer, and sometimes a Poetaster" (*Correspondence*, 186).

31. Sanborn, 154.

32. Hood, 555.

33. Clare, *Oxford Authors*, 212.

34. Thoreau, *Collected Essays and Poems*, 103.

35. Clare, *By Himself*, 100.

36. For a thoughtful discussion of the multivalent character of Clare's illness, see Bate, 408–418.

37. Thoreau, *Collected Essays and Poems*, 227.

38. Barrell, 128 and 134.

39. Clare, *Oxford Authors*, 2.

40. Thoreau, *Collected Essays and Poems*, 225.

41. Thoreau, *Collected Essays and Poems*, 229, 230, and 239.

42. Thoreau, *Collected Essays and Poems*, 224.

43. Thoreau, *Walden*, 18. For an extended meditation on the significance of crossing boundaries in Clare's life and poetry, see Goodridge and Thornton, 87–129.

44. The history of enclosure in the United Kingdom has been studied intensively. Privatization of the commons in the United States is a less familiar topic. For a historical introduction, see Montrie, *A People's History*, 13–55.

45. Bate, *John Clare: A Biography*, 219.

46. Thoreau, *Collected Essays and Poems*, 209.

47. This chapter does not engage directly with Garret Hardin's infamous argument in "The Tragedy of the Commons," which provoked a decades-long debate. For a concise critique, see Nixon.

48. Clare, *Poems of the Middle Period*, v. II, 347.

49. Clare, *Poems of the Middle Period*, v. II, 347.

50. Clare, *Poems of the Middle Period*, v. II, 347.

51. Clare, *Poems of the Middle Period*, v. II, 348.

52. Clare, *Poems of the Middle Period*, v. II, 348.

53. Clare, *Poems of the Middle Period*, v. III, 494.

54. Clare, *Poems of the Middle Period*, v. III, 495.

55. Clare, *Poems of the Middle Period*, v. II, 349.

56. Williams, 137.

57. Thoreau, *Wild Fruits*, 57–8.

58. Thoreau, *Wild Fruits*, 58.

59. Thoreau, *Wild Fruits*, 57.

60. Thoreau, *Wild Fruits*, 233.
61. Thoreau, *Wild Fruits*, 234–5.
62. Thoreau, *Wild Fruits*, 235.
63. Thoreau, *Wild Fruits*, 236.
64. Thoreau, *Wild Fruits*, 238.
65. Thoreau, *Wild Fruits*, 238.
66. This section was previously published as Newman, "Thoreau's Materialism and Environmental Justice."
67. Emerson, *Collected Works*, v. 1, 201.
68. Emerson, *Journals and Miscellaneous Notebooks*, v. 15, 353. Sherman Paul makes the definitive statement of this position.
69. Petrulionis, *Thoreau in His Own Time*. One of the first scholars to argue that we should take Thoreau's images of nature at face value was Joel Porte in *Emerson and Thoreau: Transcendentalists in Conflict* (1965). Working in part from the evidence of Thoreau's *Journal* and correspondence, he argued that Thoreau rejected Emerson's idealism (*and* his moralism) in favor of the pure aesthetics of sensory experience (91–130). According to Porte, Thoreau was in reality a passionate materialist whose "dream [was] not of some transcendent reality, but of a natural fact" (135). If Porte overstated his case when he represented Thoreau as a defiantly apolitical connoisseur of the physical textures of nature, Lawrence Buell, in *The Environmental Imagination* (1995), restored the ethical force of Thoreau's eye for detail. Buell argued that Thoreau's gradual self-education in empirical natural history led him to preservationist commitment. He identified Thoreau as the leading exemplar of a tradition of environmental literature in which "human history is implicated in natural history" and human "accountability to the environment is part of the text's ethical orientation" (7). At the same moment, Laura Walls demonstrated in *Seeing New Worlds* (1995) that Thoreau was extremely well versed in the natural philosophy of his day. He should be viewed, she argued, as an "empirical naturalist [who] saw his task to be the joining of poetry, philosophy, and science into a harmonized whole that emerged from the interconnected details of particular natural facts" (4).
70. The best synthetic account of Thoreau's intellectual development and of the organic connection between his scientific and political ideas comes in Laura Walls's magisterial new biography, *Henry David Thoreau: A Life*.
71. Emerson, *Collected Works*, v. 1, 201.
72. Thoreau, *Walden*, 4, my emphasis.
73. Thoreau, *Walden*, 5.
74. Thoreau, *Walden*, 6.

75. Thoreau, *Walden*, 6.
76. Thoreau, *Walden*, 7.
77. Thoreau, *Walden*, 8, 11.
78. Thoreau, *Walden*, 12. This truism occurs at least as early as the *Nichomachean Ethics*, where Aristotle observes that "our nature is not self-sufficient for the purpose of contemplation, but our body also must be healthy and must have food and other attention." At present, Abraham Maslow's hierarchy of needs enjoys the status of common knowledge.
79. Thoreau, *Walden*, 21.
80. Thoreau, *Walden*, 21, 26–7.
81. Thoreau, *Walden*, 28.
82. Thoreau, *Walden*, 35.
83. Thoreau, *Walden*, 91.
84. Thoreau, *Walden*, 33.
85. Thoreau, *Walden*, 8.
86. Thoreau, *Walden*, 12.
87. Thoreau, *Walden*, 13, emphasis added.
88. Thoreau, *Walden*, 12.
89. Thoreau, *Walden*, 14.
90. Schneider, 44–60, argues that Thoreau's essay uncritically reproduces Guyot's ideas in what amounts to a celebration of Manifest Destiny. On the other hand, John S. Pipkin maintains that in "Walking," Thoreau "explicitly distances himself from some aspects of Guyot's thought" and that he particularly "resists the rigid racial categories found in Guyot" (532–533).
91. Guyot, 239. Guyot mainly restated ideas that had been published two generations earlier by English and colonial American climactic determinists like Adam Ferguson and William Robertson. Thoreau does not appear to have read Ferguson's *Essay on the History of Civil Society* (1767) or Robertson's *History of America* (1777), though both were published in multiple U.S. editions during his lifetime.
92. Guyot, 247.
93. Guyot, 247–248.
94. Guyot, 77.
95. Thoreau, *Journal*, vol. 3, 182–3.
96. Thoreau, *Walden*, 15.
97. Thoreau, *Walden*, 13.
98. Thoreau, *Walden*, 37.
99. Thoreau, *Walden*, 35, emphasis in original.
100. Thoreau, *Walden*, 35.
101. Guyot, 246.
102. Thoreau, *Walden*, 61, emphasis added.

103. Thoreau, *Walden*, 56.
104. Thoreau, *Walden*, 10 and 63.
105. Thoreau, *Walden*, 30.
106. Thoreau, *Walden*, 40, emphasis added.
107. Thoreau, *Walden*, 15.
108. Thoreau, *Walden*, 90.
109. Marx, 242–65.
110. Thoreau, *Walden*, 35.
111. Thoreau, *Walden*, 53.
112. Thoreau, *Walden*, 115.
113. Thoreau, *Walden*, 116.
114. Thoreau, *Walden*, 53.
115. Thoreau, *Walden*, 92. Thoreau is punning on the British word for railroad ties: "sleepers."
116. Thoreau, *Walden*, 116.
117. Thoreau, *Walden*, 92.
118. Thoreau, *Walden*, 118. *Atropos*, one of the three fates in Greek mythology, cut the thread of each mortal's life.
119. Finley, "'Justice in the Land,'" 5.
120. Neely, 41–55.
121. Petrulionis, *To Set This World Right*, 92–5.
122. Lemire, 1–14.

BIBLIOGRAPHY

Barrell, John. 1988. *Poetry, Language, and Politics*. Manchester: Manchester University Press.

Bate, Jonathan. 1991. *Romantic Ecology: Wordsworth and the Environmental Tradition*. London: Routledge.

———. 2003. *John Clare: A Biography*. New York: Farrar, Strauss, and Giroux.

Buell, Lawrence. 1995. *The Environmental Imagination: Thoreau, Nature Writing, and the Formation of American Culture*. Cambridge, MA: The Belknap Press of Harvard University Press.

Clare, John. 1984. *John Clare: The Oxford Authors*, ed. Eric Robinson and David Powell. Oxford: Oxford University Press.

———. 1996. *Poems of the Middle Period, 1822–1837*, ed. Eric Robinson et al., 5 vols. Oxford: Clarendon Press.

———. 2002. *John Clare by Himself*, ed. Eric Robinson and David Powell. Oxford: Taylor and Francis.

Dorman, Robert L. 1998. *A Word For Nature: Four Pioneering Environmental Advocates, 1845–1913*. Chapel Hill: University of North Carolina Press.

Emerson, Ralph Waldo. 1971–. *Collected Works of Ralph Waldo Emerson*, ed. Alfred R. Ferguson, Robert E. Spiller, et al., 10 vols. to date. Cambridge, MA: The Belknap Press of Harvard University Press.

———. 1982. *The Journals and Miscellaneous Notebooks*, ed. Linda Allardt and David W. Hill, 16 vols. Cambridge, MA: The Belknap Press of Harvard University Press.

Finley, James. 2013. 'Justice in the Land': Ecological Protest in Henry David Thoreau's Antislavery Essays. *The Concord Saunterer: A Journal of Thoreau Studies*, New Series 21: 1–35.

Goodridge, John, and Kelsey Thornton. 1994. John Clare: The Trespasser. In *John Clare in Context*, ed. Hugh Haughton et al., 87–129. Cambridge: Cambridge University Press.

Guggenheim, Davis, dir. 2006. *An Inconvenient Truth*.

Guyot, Arnold. 1849. *The Earth and Man: Lectures on Comparative Physical Geography in Its Relation to the History of Mankind*. Trans. C.C. Felton. Boston: Gould, Kendall, and Lincoln. Internet Archive.

Hardin, Garrett. 1968. The Tragedy of the Commons. *Science* 162 (3859): 1243–1248.

Hood, Thomas. 1855. *Hood's Own: Or, Laughter from Year to Year*. London: Edward Moxon. *HathiTrust*.

Lemire, Elise. 2009. *Black Walden: Slavery and Its Aftermath in Concord, Massachusetts*. Philadelphia: University of Pennsylvania Press.

Lowenthal, David. 2000. *George Perkins Marsh: Prophet of Conservation*. Seattle: University of Washington Press.

Marsh, George Perkins. 1867. *Man and Nature*. New York: C. Scribner. Internet Archive.

Marx, Leo. 1964. *The Machine in the Garden*. New York: Oxford University Press.

McKusick, James. 2000. *Green Writing: Romanticism and Ecology*. New York: St. Martin's Press.

Montrie, Chad. 2011. *A People's History of Environmentalism in the United States*. London: Continuum.

Neely, Michelle C. 2013. Embodied Politics: Antebellum Vegetarianism and the Dietary Economy of Walden. *American Literature* 85 (1): 33–60.

Newman, Lance. 2015. John Clare, Henry David Thoreau, and Walking. *John Clare Society Journal* 34: 51–62.

———. 2016. Thoreau's Materialism and Environmental Justice. In *Thoreau at 200: Essays and Reassessments*, ed. Kristin Case and K.P. Van Englen, 17–30. New York: Cambridge University Press.

Nixon, Rob. 2012. Neoliberalism, Genre, and 'The Tragedy of the Commons'. *PMLA* 127 (3): 593–599.

Paul, Sherman. 1958. *The Shores of America: Thoreau's Inward Exploration*. Chicago: University of Illinois Press.

Pawelski, James, and D.J. Moores, eds. 2014. *The Eudaimonic Turn: Wellbeing in Literary Studies.* Lanham: Fairleigh Dickinson.

Peck, H. Daniel. 2005. Unlikely Kindred Spirits: A New Vision of Landscape in the Works of Henry David Thoreau and Asher B. Durand. *American Literary History* 17 (4): 687–713.

Petrulionis, Sandra Harbert. 2006. *To Set This World Right: The Antislavery Movement in Thoreau's Concord.* Ithaca: Cornell University Press.

———. 2012. *Thoreau in His Own Time.* Iowa City: University of Iowa Press.

Pipkin, John S. 2001. Thoreau's Geographies. *Annals of the Association of American Geographers* 91 (3): 527–545.

Poetzsch, Markus. 2017. The Ornithographies of John Clare and Henry David Thoreau. In *Transatlantic Literary Ecologies: Nature and Culture in the Nineteenth-Century Anglophone Atlantic World*, ed. Kevin Hutchings and John Miller, 91–104. London: Routledge.

Porte, Joel. 1966. *Emerson and Thoreau: Transcendentalists in Conflict.* Middletown: Wesleyan University Press.

Rayment, Nigel. 1990. Empiricism and the Nature Tradition. Unpublished Doctoral Dissertation, Loughborough University of Technology.

Sanborn, Franklin B. 1917. *The Life of Henry David Thoreau.* Boston: Houghton Mifflin.

Schneider, Richard, ed. 2000. *Thoreau's Sense of Place: Essays in American Environmental Writing.* Iowa City: University of Iowa Press.

Thoreau, Henry David. 1958. *The Correspondence of Henry David Thoreau*, ed. Walter Harding and Carl Bode. New York: New York University Press.

———. 1971. *Walden*, ed. J. Lyndon Shanley. Princeton: Princeton University Press.

———. 1990. *Journal*, ed. John C. Broderick, vol. 3. Princeton: Princeton University Press.

———. 2000. *Wild Fruits*, ed. Bradley Dean. New York: Norton.

———. 2001. *Collected Essays and Poems*, ed. Elizabeth Hall Witherell. New York: Literary Classics of the United States.

Walls, Laura Dassow. 1995. *Seeing New Worlds: Henry David Thoreau and Nineteenth-Century Natural Science.* Madison: University of Wisconsin Press.

———. 2017. *Henry David Thoreau: A Life.* Chicago: University of Chicago Press.

Williams, Raymond. 1973. *The Country and the City.* New York: Oxford University Press.

Afterword

Human history is a physical-linguistic process in which large-brained mammals use language and labor to shape nature into bread, sculptures, fences, cats, songs, T-shirts, tulips, books, wine, vaccines, laptops, warheads, drones, and the cloud. From a certain perspective, humans have been spectacularly successful. Around the planet, the beginning of the Cenozoic era is marked in the geologic column by a layer of iridium dust that settled from the atmosphere after a large asteroid hit what is now the Yucatán Peninsula. The ensuing impact winter snuffed out three-quarters of Earth's species. Most ecosystems were permanently altered, and mammals replaced dinosaurs as the dominant class of animals. Similarly, the beginning of the newly designated Anthropocene period is marked by a global horizon, but this one consists of fallout from nuclear explosions and particulate matter from fossil fuel combustion. In addition, there are beds of granular plastic, sills of toxic heavy metals, dikes of concrete, veins of asphalt, and other anthropogenic deposits that will eventually metamorphose into geologic features we cannot yet imagine. That is to say, human mixing of ideas with matter has left what may turn out to be one of the most prominent boundaries in the rock record of the planet's history. The paleobiologists of the future will observe a fundamental biotic shift across this boundary since current rates of species extinction are comparable to what occurred during the Cretaceous-Paleogene transition at the beginning of the Cenozoic.

© The Author(s) 2019   221
L. Newman, *The Literary Heritage of the Environmental Justice Movement*, Literatures, Cultures, and the Environment,
https://doi.org/10.1007/978-3-030-14572-9

Environmental humanists hope to do more than bear witness to what Rob Nixon calls the "slow violence" of the Anthropocene.[1] We hope to be not just subjects of this history but agents of change. For the past decade, many of us have worked under the rubric of "new materialism," trying to understand how culture, including literature, plays a part in changing the intra-actions between human societies and their material environments.[2] Rather than view nature as the ultimate reality or language as the sole source of experience, new materialists see the world as "storied matter."[3] In *Bodily Natures*, Stacy Alaimo imagines "human corporeality as trans-corporeality, in which the human is always intermeshed with [a] more-than-human world" where matter and meaning shape each other. Rather than imagine humans as spiritual/cognitive beings who once lived in tune with Nature and might do so again, the new materialism "reveals the interchanges and interconnections between various bodily natures" and "opens up a mobile space that acknowledges the often unpredictable and unwanted actions of human bodies, nonhuman creatures, ecological systems, chemical agents, and other actors."[4] Instead of a corrupted world that has fallen from its pristine original state, we see a patchy, tangled, trafficky planet that living beings have always shaped, though people do so now with much more powerful tools than have ever been available before. Some of the most powerful tools that we use to shape our world are ideas, words, and images, which are simultaneously immaterial and physical.[5] Our brains produce language through biochemical reactions that combine concepts into new ideas and then direct the instrument of the larynx to vibrate the gaseous atmosphere in order to communicate them to other brains. The primary function of the material processes of thought, speech, and writing is to coordinate our collective labor. This is not just a matter of naming objects, making plans, and giving orders. Like a hoe, a *topos* is a tool, a contrivance that we use, like Henry Thoreau, to make the Earth say beans or revolution.[6]

In addition to offering an integrative way of thinking about nature and culture, new materialism imagines a new ethical orientation that "denies the human subject the sovereign, central position" and conceives of "matter as possessing its own modes of self-transformation, self-organization, and directedness."[7] In other words, all material objects are purposeful (if not necessarily self-aware) beings whose self-directed otherness confers ethical standing. As we move through the material-semiotic space that Timothy Morton calls the "mesh," we encounter nonhuman others (including animals, plants, soils, mountains, rivers, breezes, and floods of light) as "strange strangers" who are moving at their own paces, pursuing their own goals, and changing the world in their own ways. They deserve our respect and consideration. Because new materialism reorients environmental ethics in this way, it has the potential to ground the global environmental justice movement. In Ursula Heise's terms, the new materialism articulates "a sense of planet" that makes it possible for us to live as "eco-cosmopolitan" members of a planetary community.[8]

In order to fully achieve this potential, though, new materialism will need to walk a few steps further. So far, it has mainly imagined humans as solitary undifferentiated individuals traversing the more-than-human world. But, of course, we are social beings. We live in the world not as individually embodied minds, not as free radicals, but as embedded members of an ecosocial body that has been (dis)organized by global capitalism along lines of class, race, gender, sexuality, and even species. Evolving our way out of the Anthropocene will require us to recognize and practice ethical commitments across these differences too. New materialism can best help us do so by studying how different people have intra-acted with and written about the storied landscapes we live in. Some of the most useful insights may come from studying those who invented news ways of writing about nature in order to communicate the urgency of their struggles for social and environmental justice.

The Environmental Heritage of the Environmental Justice Movement has been an experiment in new materialist literary history. My main interpretive method has been sustained close reading of primary texts. My main goal has been to draw attention to powerful *topoi* that speak to the concerns of new materialists and the environmental justice movements they serve. While I do not open a sustained theoretical inquiry, I hope that the texts showcased here raise new generative questions. How can our environmental ethics acknowledge that we interact with nature in ways that are shaped by our identities and positions? How do classed, raced, and gendered bodies differently experience the material flows and accumulations of modernity? How are our ecosocial identities and histories differently mediated by environmental literature and discourse?[9] How might our identities and positions support our work as agents of change? How can we reframe the images of nature and disrupt the narratives of conquest that capitalism has used to justify its conquest of the planet? Can we tell stories that will convoke new forms of solidarity and alliance? Can we create visions that help new communities of resistance see their way to a just green future?

In "Book VII" of *The Prelude, or, The Growth of a Poet's Mind*, William Wordsworth claims that, after leaving Cambridge, he feels "Well pleased to pitch a vagrant tent among/The unfenced regions of society."[10] Despite his bland confidence about visiting the urban wilderness of London, he is dazed by "the shock/ Of the huge town's first presence."[11] The city appears as a "monstrous ant-hill on the plain/Of a too-busy world." "Chapter VII" of *The Prelude* takes the form of a pedestrian excursion on the city's streets, and the poet is awestruck by the "comers and the goers face to face,/Face after face." Human individuality is erased in an "endless flow of men and moving things."[12] The poet perceives "the thickening hubbub" of people, not as a series of distinct beings, but as a parade of types. He catalogs passers-by according to their nationality (Italian, Jew, Turk), their occupation (nurse, idler, scavenger), and "the colors which the sun bestows":

> The Swede, the Russian; from the genial south,
> The Frenchman and the Spaniard; from remote
> America, the Hunter-Indian; Moors,
> Malays, Lascars, the Tartar, the Chinese,
> And Negro Ladies in white muslin gowns.[13]

The teeming streets of London present a vertiginous spectacle of categories. At Bartholomew Fair, which was a mercantile exchange with popular midway entertainments, the crowd merges into a "huge fermenting mass of human-kind."[14] In response to the "anarchy and din,/Barbarian and infernal," the poet dissociates himself from the imagined scene by means of a *deus ex machina*: "the Muse's help will we implore,/and she shall lodge us, wafted on her wings,/Above the press and danger of the crowd."[15] Like Ralph Waldo Emerson floating above Boston Common, Wordsworth looks down on the midway, and it is "alive with heads." This "Parliament of Monsters" is "a true epitome/of what the Mighty City is herself."[16] Wordsworth's London is a synecdoche of modernity, an emblem of a nightmarish world in which globalization under capitalism has erased traditional social hierarchies, leaving only a planetary mob that threatens to spill out of the city and overrun the quiet, clean English countryside.

The first U.S. edition of *The Prelude* was published in 1850 at 200 Broadway in New York City by D. Appleton and Company. At that time, the young Walt Whitman was leaving behind his career as a political journalist to become a poet. In May 1851, Whitman read a review of *The Prelude*, which is subtitled "An Autobiographical Poem," and he almost certainly read the poem itself at about the same time, probably in the small bookstore he ran in Brooklyn. His own autobiographical poem,

"Song of Myself," was published just four years later. It is a working-class writer's response to Wordsworth's *magnum opus*. Like *The Prelude*, "Song of Myself" takes the form of a pedestrian excursion, and it centrally features a cityscape crowded with human bodies. But instead of recoiling at proximity to difference, Whitman's poet-prophet asks, "Who need be afraid of the merge?"[17] He celebrates intimacy with the global community that assembles and mixes on the streets: "Of every hue and trade and rank, of every caste and religion,/Not merely of the New World but of Africa Europe or Asia." He enthusiastically inhabits and even embodies this leveled social world: "I resist anything better than my own diversity,/And breathe the air and leave plenty after me,/And am not stuck up, and am in my place."[18] When "Song of Myself" appeared, the United States was moving quickly toward civil war. The poem rejects the white supremacist ideology of the slave South, and it offers New York City as a harbinger of a multicultural, democratic nation to come. It unfurls a documentary panorama of people of all kinds working side by side, and it lingers occasionally to spotlight a representative individual:

> The negro that drives the huge dray of the stoneyard . . . . steady and tall he
>   stands poised on one leg on the stringpiece,
> His blue shirt exposes his ample neck and breast and loosens over his hipband,
> His glance is calm and commanding . . . . he tosses the slouch of his hat away
>   from his forehead,
> The sun falls on his crispy hair and moustache . . . . falls on the black of his
>   polish'd and perfect limbs.[19]

Not only do human beings of all kinds merge into a spectacle of cooperative vitality, but the country and the city merge into an integrated landscape of free streets and free soil:

> The crew of the fish-smack pack repeated layers of halibut in the hold,
> The Missourian crosses the plains toting his wares and his cattle,
> As the fare-collector goes through the train – he gives notice by the jingling
>   of loose change,
> The floor-men are laying the floor – the tinners are tinning the roof – the
>   masons are calling for mortar,
> In single file each shouldering his hod pass onward the laborers;
> Seasons pursuing each other the indescribable crowd is gathered ...
> Seasons pursuing each other the plougher ploughs, the mower mows, and
>   the wintergrain falls in the ground.[20]

The continent is not differentiated into the usual categories: rural and urban, wild and tame, pristine and degraded, healthy and sick. Instead, the land is pictured as nurturing habitat for an integrated human community that subsists through shared labor. The thriving multitude that the poem gathers in imagination is a performative assembly that lays claim to the national commons for its own use and care: "It is for the endless races of working people."[21] The spirit of solidarity that pervades this utopian eco-social community is inspired by people's shared physicality: "For every atom belonging to me as good belongs to you."[22] The "leaves of grass" of the poem's title are the people of the assembly on the commons: "This is the grass that grows wherever the land is and the water is,/This the common air that bathes the globe."[23] Walking through the green grass, Whitman reveals humanity's shared roots in the material body of the earth, and he envisions a future in which people love one another and their shared home.

Notes

1. Nixon, n.p.
2. The word "materialism" has several distinct but related uses. In philosophical discourse, materialism is opposed to idealism and refers to the ontological position that the world is composed only of matter and its motions. This sense has roots at least as far back as the writings of Democritus, Epicurus, and Lucretius. This kind of materialism is often conflated with or accompanied by empiricism, the epistemological position that our ideas about the nature of things are formed exclusively by sensory experience of matter and its motions. Empiricism, of course, also has a long history, especially since it was formalized by Locke and Hume. In common usage, on the other hand, materialism usually means the tendency to value physical (and sometimes financial) possessions and interests, over and above spiritual, aesthetic, and intellectual experiences and concerns. Finally, in the context of political theory, materialism names the belief that structures of economic power, rather than (or in addition to) ideas or ideology, determine (or shape) social being and history. The common thread here, of course, is a set of familiar binary oppositions: mind and matter, spirit and nature, words and things. In the last decade or two, in the context of the environmental humanities, materialism has been used to signal a turn away from the reductive linguistic determinism that has dominated academic discourse for the last several decades. It is only in this context that it makes any sense at all to call materialism new. New materialists in the environmental humanities state their motivations in remarkably consistent terms.

If we acknowledge that everything including ourselves is material, we will flatten the ethical field in a way that makes new ethical commitments possible. If we think differently about matter, we will act differently toward it. If we acknowledge that we are material beings living in a material world, we will treat our home with care. Key statements of the new materialism include Alaimo, Barad, Bennett, Coole and Frost, Dolphijn and van der Tuin, and Morton.

3. Iovino and Oppermann, "Theorizing Material Ecocriticism," 451. Iovino and Oppermann played a particularly important role in introducing new materialism to the ecocritical community centered on the Association for the Study of Literature and the Environment. See also their article, "Material Ecocriticism."

4. Alaimo, 2.

5. Integrative thinking of this kind has been a feature of the best ecocritical theory for some time. In his contribution to the landmark *Ecocriticism Reader*, William Howarth observed that "writers and their critics are stuck with language, and although we cast *nature* and *culture* as opposites, in fact they constantly mingle, like water and soil in a flowing stream" (69). And in her introduction to the same collection, Cheryll Glotfelty observes that literature "does not float above the material world in some aesthetic ether, but, rather, plays a part in an immensely complex global system, in which energy, matter, *and ideas* interact" (xix).

6. Thoreau has inspired some new materialists. For instance, Jane Bennett meditates on his vegetarianism and suggests that he imagines eating as an "encounter" between the human body and vital foodstuffs. During that encounter, "all bodies are shown to be but temporary congealments of a materiality that is a process of becoming" (49). And Rochelle Johnson has recently argued that Thoreau can help us achieve a holistic view of matter since he "experienced an integration between the human mind, the matter of spirit, and the realm of nonhuman matter that informed – even *constituted* – his being" (624). According to Johnson, his writings not only help us understand the vitality of matter but also acknowledge the materiality of thought and experience. In particular, she argues that Thoreau demonstrates how enchantment and ineffable meaning can emerge from the intra-activity of the embodied mind (which "is shaped by – and *is* – matter") with the physical objects and processes that surround, infuse, and engender it (614).

7. Alaimo, 16; Coole and Frost, 10.

8. Heise 50–66.

9. Sarah Jaquette Ray's book, *The Ecological Other*, is a powerful study of environmental discourses of bodily difference: "Disgust shapes mainstream environmentalism and vice versa, and it does so by describing which kinds

of bodies and bodily relations to the environment are ecologically 'good,' as well as which kinds of bodies are ecologically 'other'" (1). Moreover, "environmentalism's focus on maintaining a pure nature is as much about social order as it is about ecological health" (2). Ray examines the function of the body "in discourses of risk in American outdoor adventure culture, in representations of the Native American body in environmental justice literature, and in attitudes toward the immigrant body in contemporary environmental justice discourse" (9), and she shows that from its beginnings, environmentalism has been predicated on a normative white, male, able body that was as much the target of conservation as the segregated wilderness where it performed its fitness. Ray points out that the environmental humanities' materialist turn to the body as the means of healthy connection to the environment risks reproducing many of these old and all-too-familiar exclusions, and she argues for a rigorously inclusive environmentalism within which the ecological other is not only permitted to speak but even "to revise mainstream environmentalism entirely and challenge assumptions of what 'environmental' means" (180).

10. Wordsworth, 173.
11. Wordsworth, 174.
12. Wordsworth, 177.
13. Wordsworth, 180.
14. Wordsworth, 197.
15. Wordsworth, 199.
16. Wordsworth, 201.
17. Whitman, 17.
18. Whitman, 24.
19. Whitman, 20.
20. Whitman, 23.
21. Whitman, 24.
22. Whitman, 13.
23. Whitman, 24.

BIBLIOGRAPHY

Alaimo, Stacy. 2010. *Bodily Natures: Science, Environment, and the Material Self.* Bloomington: Indiana University Press.

Barad, Karen. 2007. *Meeting the Universe Halfway: Quantum Physics and the Entanglement of Matter and Meaning.* Durham: Duke University Press.

Bennett, Jane. 2010. *Vibrant Matter: A Political Ecology of Things.* Durham: Duke University Press.

Coole, Diana H., and Samantha Frost. 2010. *New Materialisms: Ontology, Agency, and Politics.* Durham: Duke University Press.

Dolphijn, Rick, and Iris van der Tuin. 2012. *New Materialism: Interviews & Cartographies.* Ann Arbor: Open Humanities Press.

Glotfelty, Cheryl, and Harold Fromm, eds. 1996. *The Ecocriticism Reader.* Athens: University of Georgia Press.

Heise, Ursula. 2008. *Sense of Place and Sense of Planet: The Environmental Imagination of the Global.* New York: Oxford University Press.

Howarth, William. Some Principles of Ecocriticism. In The Ecocriticism Reader, Cheryl Glotfelty and Harold Fromm, eds. 69–91.

Iovino, Serenella, and Serpil Oppermann. 2012a. Material Ecocriticism: Materiality, Agency, and Models of Narrativity. *Ecozon@* 3 (1): 75–91.

Iovino, Serenella, and Serpil Oppermann. 2012b. Theorizing Material Ecocriticism. *ISLE: Interdisciplinary Studies in Literature and the Environment* 19 (3): 448–475.

Johnson, Rochelle. 2014. 'This Enchantment Is No Delusion': Henry David Thoreau, the New Materialisms, and Ineffable Materiality. *ISLE* 21 (3): 606–635.

Morton, Timothy. 2010. *The Ecological Thought.* Cambridge, MA: Harvard University Press.

Nixon, Rob. 2011. *Slow Violence and the Environmentalism of the Poor.* Cambridge, MA: Harvard University Press.

Ray, Sarah Jaquette. 2013. *The Ecological Other: Environmental Exclusion in American Culture.* Tucson: University of Arizona Press.

Whitman, Walt. 1855. *Leaves of Grass.* New York. *The Walt Whitman Archive.*

Wordsworth, William. 1850. *The Prelude; or, Growth of a Poet's Mind.* New York: D. Appleton and Co. *HathiTrust.*

INDEX

© The Author(s) 2019
L. Newman, *The Literary Heritage of the Environmental Justice
Movement*, Literatures, Cultures, and the Environment,
https://Doi.org/10.1007/978-3-030-14572-9

9 783030 145712